IMMOBILIZED ENZYME TECHNOLOGY

Research and Applications

IMMOBILIZED ENZYME TECHNOLOGY
Research and Applications

Edited by

Howard H. Weetall
Research and Development Laboratory
Corning Glass Works
Corning, New York

and

Shuichi Suzuki
Research Laboratory of Resources Utilization
Tokyo Institute of Technology
Tokyo, Japan

PLENUM PRESS • NEW YORK AND LONDON

Library of Congress Cataloging in Publication Data

Seminar on Research and Development of Immobilized Enzymes, Tokyo, 1974.
   Immobilized enzyme technology.

   Sponsored jointly by the National Science Foundation and the Japan Society
for the Promotion of Science.
   Includes bibliographical references and index.
   1. Immobilized enzymes—Industrial applications—Congresses. I. Weetall, Howard
H. II. Suzuki, Shuichi, 1924-   III. United States-Japan Cooperative Science
Program. IV. United States. National Science Foundation. V. Nippon Gakujutsu
Shinkōkai. VI. Title.
TP248.E5S45   1974   660'.63   75-17858

ISBN -13 : 978-1-4615-8737-8   e-ISBN-13 : 978-1-4615-8735-4
DOI : 10.1007/978-1-4615-8735-4

Softcover reprint of the hardcover 1st edition 1975

Proceedings of The U.S.–Japan Cooperative Science Program Seminar held
in Tokyo, Japan, November 5-8, 1974

© 1975 Plenum Press, New York
A Division of Plenum Publishing Corporation
227 West 17th Street, New York, N.Y. 10011

United Kingdom edition published by Plenum Press, London
A Division of Plenum Publishing Company, Ltd.
Davis House (4th Floor), 8 Scrubs Lane, Harlesden, London, NW10 6SE, England

Preface

On November 5-8, 1974, a Seminar on Research and Development of Immobilized Enzymes was held in Tokyo, Japan. The seminar was part of the United States-Japan Cooperative Science Program sponsored jointly by the National Science Foundation and the Japan Society for the Promotion of Science.

The purpose of the seminar was to promote the scientific exchange of ideas and scientific results of a practical nature, as well as academic advances made in both countries through discussion and exchange of ideas.

The areas chosen for discussion included: fundamental research in immobilized enzymes, new techniques of enzyme purification, comparative studies on immobilization techniques and the relative merits, chemical engineering aspects of the technology, industrial applications and reactor design.

The discussions and exchange of ideas which took place at the seminar should promote new research and development which we hope will lead to new important advances in enzyme technology.

These proceedings represent the summation of the work presented and discussed at the seminar. The editors hope the reader will find them interesting and informative.

Howard H. Weetall
Shuichi Suzuki
March, 1975

Contents

IMMOBILIZATION OF COENZYME B$_6$ AND SEVERAL B$_6$ ENZYMES: APPLICATION TO ASSAY OR PRODUCTION OF SOME AMINO ACIDS AND TO STRUCTURE-FUNCTION INTERRELATIONSHIP STUDIES OF B$_6$ ENZYMES

Saburo Fukui and Sei-ichiro Ikeda

Laboratory of Industrial Biochemistry, Department of Industrial Chemistry, Faculty of Engineering, Kyoto University, Kyoto, Japan

Although a number of papers have been published concerning immobilization of various enzymes, little has been reported with respect to immobilization of enzymes requiring coenzymes. This paper describes immobilization of coenzyme B$_6$ (pyridoxal 5'-phosphate, PLP) and several B$_6$ enzymes which require PLP as coenzyme and catalyze a variety of reactions of amino acids. Applications of the immobilized B$_6$ enzymes to production or assay of some amino acids and to studies on the structure-function interrelationship are comprehensively mentioned.

(I) PREPARATION OF SEPHAROSE-BOUND PYRIDOXAL 5'-PHOSPHATE

Three types of Sepharose-bound PLP were newly prepared. SP-A (6-immobilized PLP) was obtained by coupling diazotized p-aminobenzamido-hexyl-Sepharose to 6-position of PLP as reported previously (1). Other two Sepharose-bound PLP, SP-B (N-immobilized PLP) and SP-C (3-0-immobilized PLP) were prepared by reacting PLP with a bromoacetyl derivative of Sepharose 4B in dimethylformamide (50%, v/v) and in potassium phosphate buffer (pH 6.5), respectively (2) (Fig. 1). The Sepharose-bound PLP derivatives thus obtained were characterized by several lines of experiments. SP-A has all of the functional groups of PLP necessary for the appearance of catalytic activity in both enzymatic and non-enzymatic reactions. Treatment of SP-A with sodium dithionite in 0.2 M sodium borate buffer resulted in liberation of a pyridoxine-like compound having an absorption maximum at ca. 316 nm at pH 9.0 and at 282 nm at pH 3.0, respectively. From the optical density of this compound (probably,

1

6-aminopyridoxine 5'-phosphate), it was estimated that SP-A contained ca. 1.5 μ moles of PLP per gram of Sepharose. On the other hand, the catalytic activity of SP-A in the non-enzymatic cleavage of tryptophan (3) corresponded to ca. 1.5 - 3.0 μ moles of PLP per gram of Sepharose. SP-B and SP-C showed an absorption maximum at 295 nm in 0.1 M NaOH. In 0.1 M NaOH, SP-B showed an absorbance maximum at 388 nm observed in free PLP or N-methyl PLP, while SP-C showed an absorption peak at 315-335 nm as observed in 3-0-methyl PLP. The PLP contents of SP-B and SP-C were calculated from the absorbance at 295 nm in 0.1 M HCl. More clear evidence for the structures of SP-B and SP-C was obtained by the absorption spectra of the substrate liberated from the gels previously reduced with NaBH₄ by treatment with 6 M HCl. The eluate from the reduced SP-B showed absorption spectra similar to those of N-methylpyridoxine at various pHs, and the eluate from the reduced SP-C exhibited absorption spectra analogous to those of 3-0-methylpyridoxine at different pHs.

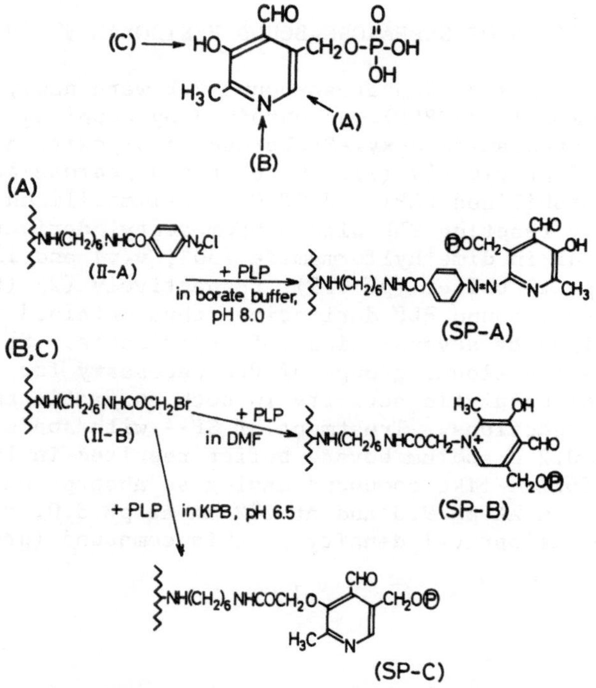

Fig. 1: Preparation of Sepharose-bound PLP

TABLE 1

Characterization of three kinds of Sepharose-bound PLP

Sepharose-bound PLP derivative	Immobilized position (in PLP)	PLP content[1] (moles PLP/ g Sepharose)	Activity in non-enzymatic catalysis[2] (moles PLP/g Sepharose)	Affinity for apo-Trpase[3] etc.
SP-A	6	1.5	1.5 - 3.0	+
SP-B	1-N	1.2	0.8 - 1.1	+
SP-C	3-0	0.84	0 - trace	+

[1] Calculated from spectral data.
[2] Calculated from the catalytic activities in the non-enzymatic cleavage of trypotphan.(3)
[3] Judged from the results in affinity chromatography and immobilization of B$_6$ enzymes.

Table 1 summarizes the characterization of three kinds of Sepharose-bound PLP. SP-B served as a catalyst for the non-enzymatic tryptophan cleavage. SP-C, however, did not exhibit any appreciable catalytic activity, as expected from the well-established essential role of 3-OH of PLP. All of these immobilized PLP had good affinities for apoproteins of various B$_6$ enzymes. For example, SP-C is useful for the affinity chromatography of tryptophanase and β-tyrosinase.

(II) IMMOBILIZATION OF SEVERAL B$_6$ ENZYMES BY USE OF SEPHAROSE-BOUND PYRIDOXAL 5,-PHOSPHATE OR DIRECTLY ON SEPHAROSE

Immobilization of several B$_6$ enzymes was carried out by three different methods using Sepharose 4B as insoluble carrier. The enzymes tested were tryptophanase from E. coli B/1t 7-A (4-6), tyrosine phenol-lyase (β-tyrosinase) from E. intermedia (7,8), aspartate 4-decarboxylase from Pseudomonas dacunhae (9,10) and aspartate aminotransferase from pig heart (11,12).

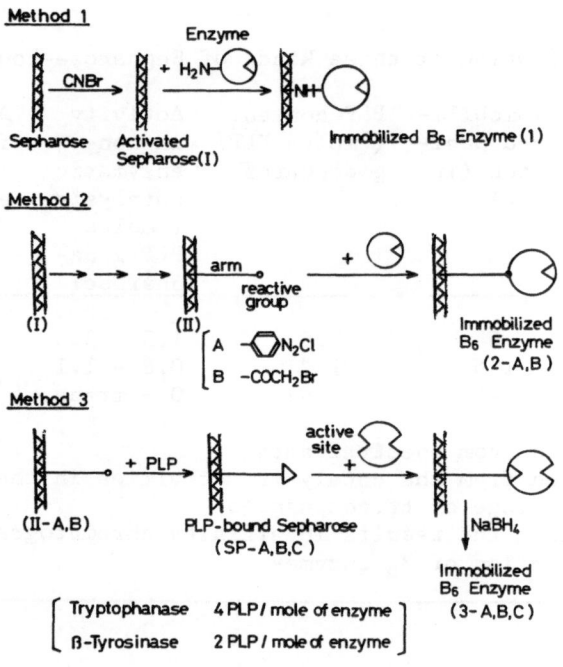

<u>Scheme I.</u> Procedure for Immobilization.

The principle of the immobilization procedures are illustrated in Scheme I. Method 1: Enzymes were immobilized
directly on CNBr-activated Sepharose 4B (Scheme I, Method
1); Method 2: Enzymes were immobilized on Sepharose through
a reactive side arm. A diazonium derivative of Sepharose
(Scheme I, Method II-A) and a bromoacetyl derivative of
Sepharose (Scheme I, Method II-B) were used for the immobilization. The detailed description has been published elsewhere (1,13,14). Method 3: Enzymes were immobilized on
Sepharose through PLP previously bound to Sepharose. Three
kinds of Sepharose-bound PLP mentioned above were used for
the purpose. For example, the immobilization of apo-trypto-
phanase (Trpase) on SP-A was carried out as follows: Wet
SP-A (0.5 g) was mixed with 1.0 ml of 0.1 M potassium phosphate buffer (pH 7.0) containing 0.50 mg of apo-Trpase. The
mixture was incubated for 20 min at 37°C. The resulting
SP-A-apo-Trpase complex was reduced with NaBH₄ and then
washed thoroughly with a mixture of 0.1 M potassium phosphate

Table 2. Comparison of immobilization efficiency and enzyme
          activity of tryptophanase and β-tyrosinase prepared
          by different methods.

Expt. No.	Method[1]	Sepharose der. used (g)	Enzyme form & Amts. used (mg)		Immobilization efficiency[2] (%)	Relative activity[3] (%)
1. Tryptophanase						
1	1	I 0.50	0.20	(apo)	73	34
2	1	I 0.50	0.20	(holo)	68	42
3	2-A	II-A 0.50	0.50	(holo)	42	10
4	2-B	II-B 0.50	0.50	(holo)	65	35
5	3-A	SP-A 0.50	0.50	(apo)	81	60
6	3-B	SP-B 0.50	0.50	(apo)	21	51
7	3-C	SP-C 0.50	0.50	(apo)	33	48
2. β-Tyrosinase						
1	1	I 1.09	2.00	(apo)	59	41
2	1	I 1.36	5.00	(apo)	98	40
3	1	I 1.06	2.00	(holo)	60	30
4	2-B	II-B 0.44	0.50	(apo)	94	41
5	2-B	II-B 0.49	0.50	(holo)	94	33
6	3-A	SP-A 0.52	0.50	(apo)	58	29

1) Method 1: Enzyme was reacted with CNBr-activated Sepharose 4B.
   Method 2-A: Enzyme was coupled with diazotized p-aminobenzamido-
   hexyl-Sepharose.
   Method 2-B: Enzyme was coupled with bromoacetamidohexyl-Sepharose.
   Method 3: Enzyme was coupled to Sepharose-bound PLP.
2) The ratio of the amount of immobilized enzyme protein to the amount
   of enzyme protein used initially for immobilization.
3) Enzyme activity was measured by a batch method in the presence of added
   PLP and was expressed as a relative value to the specific activity of
   the respective free counterpart.

buffer (pH 6.0). 5mM PLP and 11.4% ammonium sulphate. The
third method employed first by the present authors is based
upon the specific affinity between the coenzyme and the apoenzyme at the active center. Since the Schiff base linkage
between PLP and ε-amino group of the lysine residue at an
active center of Trpase or β-tyrosinase is not sufficiently
strong to obtain a stable immobilized enzyme, the linkage
was fixed by reduction with $NaBH_4$. Table 2 shows the comparison of the efficiency of immobilization and the activity
of the immobilized Trpase and β-tyrosinase preparations obtained by the above-mentioned different methods. In the
case of Trpase having 4 subunits and 4 active centers, the
most active immobilized enzyme was obtained by the reaction
of apo-Trpase with SP-A (immobilization efficiency, 81%;
relative activity of the resulting immobilized enzyme, 60%).
In the case of β-tyrosinase having two active sites, how-

Table 3. Kinetic properties of immobilized enzymes

Enzyme preparations		Km (mM)	Kco (µM)
Tryptophanase	soluble	0.33	1.1
	immobilized	0.34	1.2
ß-Tyrosinase	soluble	0.23	1.3
	immobilized	0.27	5.0

ever, this method was not so suitable as the case of Trpase. This would be explained by the facts that one of the two catalytic sites of β-tyrosinase was blocked by the binding to Sepharose-bound PLP and the resulting distortion of enzyme conformation may be severer than the case of Trpase having tetrameric structure. The immobilization efficiency of β-tyrosinase by direct reaction with CNBr-activated

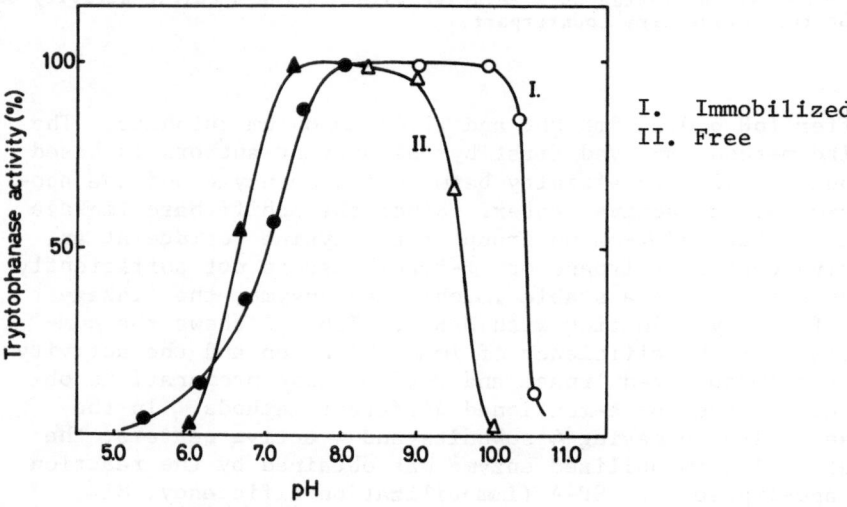

I.   Immobilized
II. Free

Fig. 2: Effect of immobilization on the pH-activity curve
          of Trpase

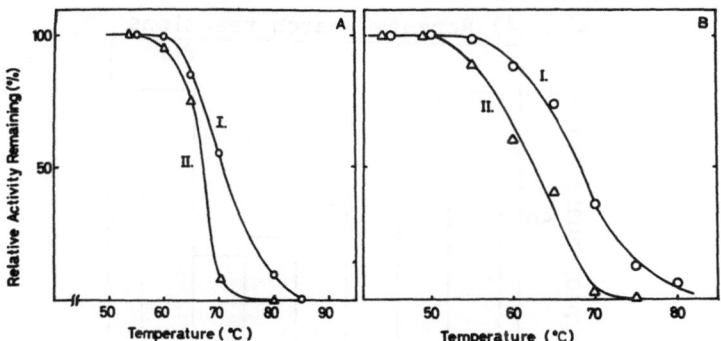

<u>Fig. 3</u>: Heat stability of immobilized Trpase and β-tyrosi-
nase.
A. Trpase;   B. β-Tyrase;   I. Immobilized;   II. Free

Sepharose or Sepharose derivatives having a suitable reactive side arm was <u>ca</u>. 90% and the resulting immobilized
preparations showed <u>ca</u>. 40% relative activities of the free
enzyme.

    Using the most active preparations of Trpase and   - β
tyrosinase, the fundamental enzymatic properties were studied in comparison with the respective free counterparts.
As shown in Table 3, kinetic properties such as affinities
for coenzyme or substrate were scarcely altered upon immobilization, suggesting that the steric environment around
the active center of each enzyme is not influenced by the
immobilization to an appreciable extent. On the contrary,
the pH optima of both enzymes shifted 0.5 to 1.0 pH unit
toward alkaline side upon immobilization. Fig. 2 shows the
change of pH-activity profile of Trpase by immobilization.
In the α,β-elimination reaction catalyzed by Trpase or   - β
tyrosinase, the elimination of α-hydrogen of the amino acid
substrate as proton has been shown to occur more rapidly
prior to the subsequent rate-determining β-elimination step.
The localization of this proton may cause the shift of the
pH optima.

    As illustrated in Fig. 3, both immobilized Trpase and
β-tyrosinase showed high thermal stability as compared with
their free counterparts. In connection with these results,

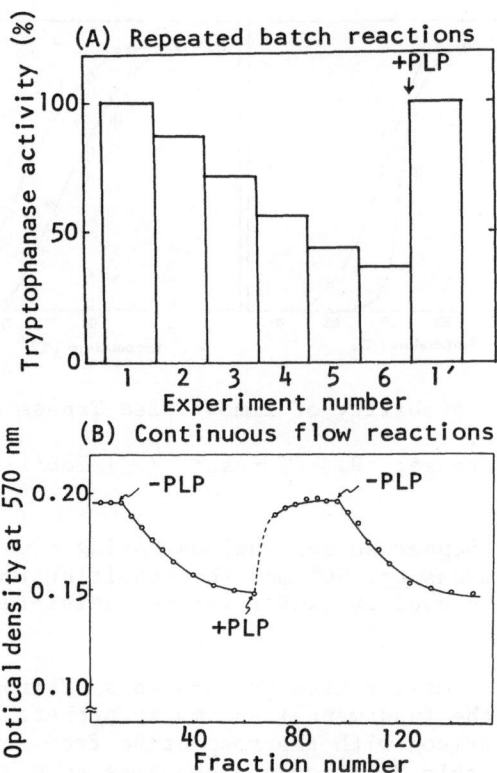

<u>Fig. 4</u>:   Decrease of activity caused by liberation of co-
enzyme.

the optimal temperature of each immobilized enzyme was
fairly higher than that of the respective free enzyme in a
short time-reaction (e.g., 10 min).

As depicted in Fig. 4, both immobilized Trpase and β-
tyrosinase reconstituted with PLP lost their activities
gradually when used repeatedly without added PLP in a batch
system or a continuous flow system. But the initial activities were restored by supplement of PLP to the reaction
systems. These results indicate that each of the enzyme
protein immobilized on Sepharose matrix would be fairly
stable and the coenzyme part is readily resolved from the
holoenzyme.

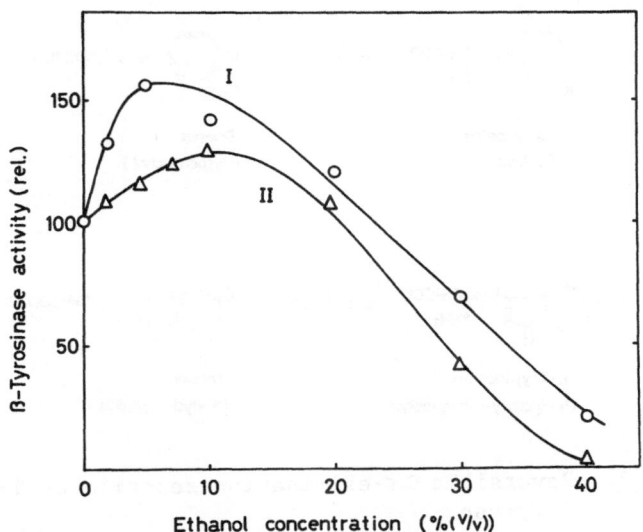

<u>Fig. 5</u>: Effect of ethanol.

Both immobilized enzymes exhibited high resistence to denaturing agents such as guanidine-HCl and to organic solvents such as ethanol. Treatment of immobilized β-tyrosinase with a low concentration of guanidine-HCl (up to 4 M) resulted in a marked enhancement of the enzyme activity. Also, as shown in Fig. 5, addition of a low concentration of ethanol stimulated the rates of both enzymatic reactions (13,14).

(III) PRODUCTION OF L-TRYPTOPHAN, L-TYROSINE AND THEIR ANALOGUES BY USE OF IMMOBILIZED TRYPTOPHANASE OR β-TYROSINASE

Recently, Yamada and his coworkers have developed effective methods for the production of L-tryptophan, L-tyrosine and their physiologically important analogues by use of soluble tryptophanase or β-tyrosinase or by use of the intact bacterial cells containing each of these enzymes abundantly. As shown in Scheme 2, L-tyrosine or 3,4-dihydroxyl-L-phenylalanine (L-DOPA) is synthesized from pyruvate, ammonia and phenol or pyrocatechol by β-tyrosinase reverse reaction. In an analogous manner, L-tryptophan or

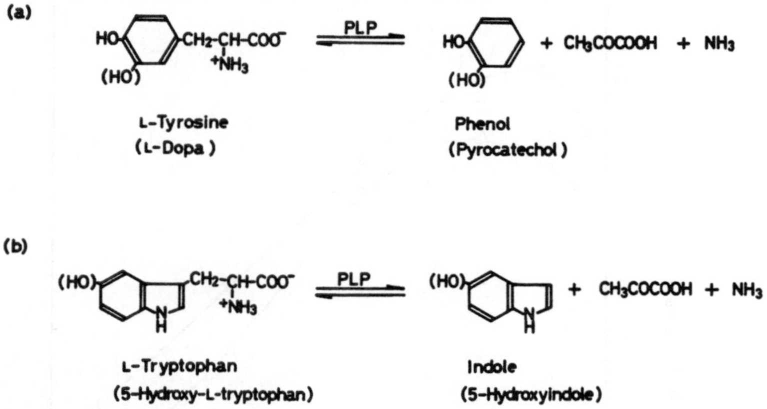

(a)

L-Tyrosine
(L-Dopa)

Phenol
(Pyrocatechol)

(b)

L-Tryptophan
(5-Hydroxy-L-tryptophan)

Indole
(5-Hydroxyindole)

Scheme 2. Reversible α,β-elimination reaction by Trpase and β-Tyrase.

5-hydroxy-L-tryptophan is synthesized from pyruvate, ammonia and indole or 5-hydroxy-indole by tryptophanase reverse reaction (15,16). Hence, we attempted the application of immobilized Trpase and of β-tyrosinase to the production of these amino acids in cooperation of Yamada et al.

As depicted in Fig. 6, L-tryptophan and its analogue were synthesized conveniently in good yield by a continuous flow method or by a repeated batch method using immobilized Trpase (13,17). For example, the reaction mixture indicated in Fig. 6 was passed through a jacketted column (1.2 x 10 cm) packed with 10 ml of immobilized Trpase at 37°C with an appropriate flow rate (e.g., SV: below 5 h^{-1}). Under these conditions, ca. over than 90% of indole used was converted to L-tryptophan. Both immobilized enzymes were fairly stable during these reactions and were useful as excellent catalysts during long-term continuous reactions.

(IV) APPLICATION OF IMMOBILIZED B_6 ENZYMES TO MICRO-ASSAYS OF AMINO ACIDS

The most common analytical method for L-tryptophan consists of condensation with a suitable aldehyde, e.g., p-dimethylaminobenzaldehyde, followed by oxidation with nitrite. In comparison with tryptophan, however, indole

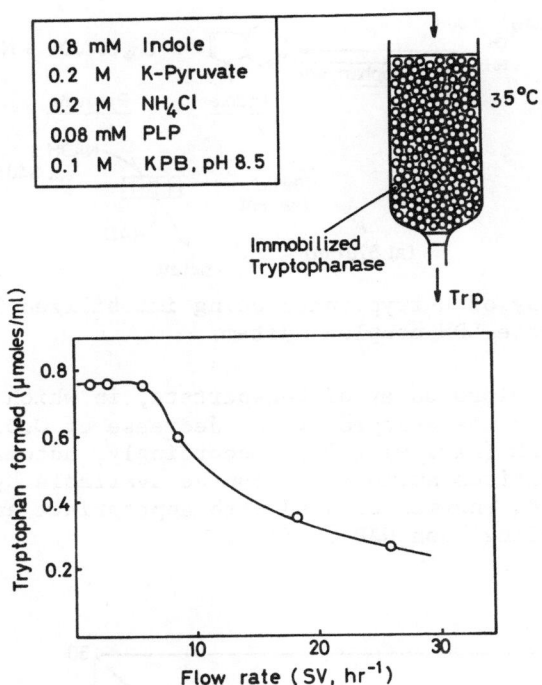

0.8 mM   Indole
0.2 M   K-Pyruvate
0.2 M   NH$_4$Cl
0.08 mM PLP
0.1 M   KPB, pH 8.5

35°C

Immobilized
Tryptophanase

Trp

Fig. 6:   Production of L-tryptophan by a continuous flow
           method using a column of immobilized Trpase.

can be more conveniently assayed with the aldehyde alone
with much higher sensitivity.   Application of immobilized
Trpase column for the conversion of L-tryptophan to indole
was very successful.   The calibration curve of L-tryptophan
treated with an appropriate immobilized Trpase column was
almost identical with that of free indole, when acid-Ehrlich
reagent was used.   Moreover, the amount of L-tryptophan applied to a column of immobilized Trpase and lactate dehydrogenase (LDH)-coupled system was quantitatively determined by
the decrease of NADH (O.D. at 430 nm) (Fig. 7 and 8) (18).
In an analogous way, L-aspartic acid was spectrophotometrically assayed after converted to oxalacetate (OAA) using a
column of immobilized aspartate aminotransferase (AAT).
Furthermore, a coupled system of AAT and malate dehydrogenase (MDH) immobilized concomitantly on Sepharose was very

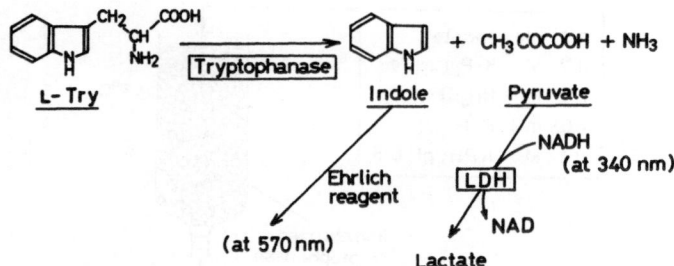

Fig. 7:   Assay of L-tryptophan using immobilized Trpase or
             Trpase-LDH coupled system.

useful for a micro-assay of L-aspartate, in which the amount
of L-aspartate was assayed by the decrease of O.D. at 340
nm due to NADH (Fig. 9) (19). Accordingly, automatic micro-
assays for various amino acids can be available by use of
immobilized B_6 enzymes coupled with appropriate enzyme sys-
tem(s) depending upon NAD or NADH.

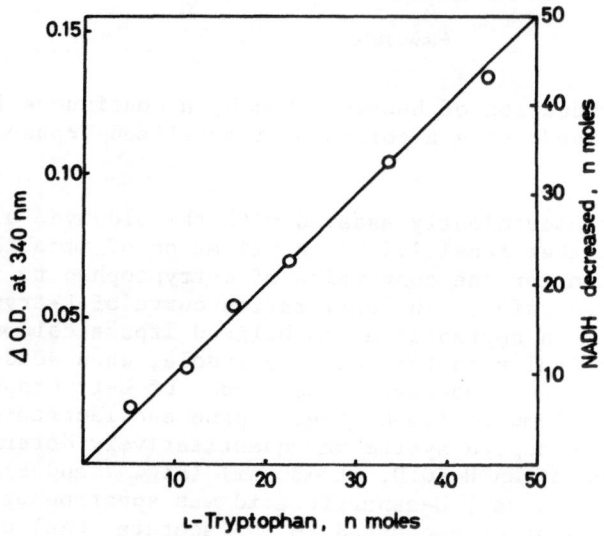

Fig. 8:   Calibration curve for L-tryptophan in Trpase-LDH
             coupled system immobilized on Sepharose.

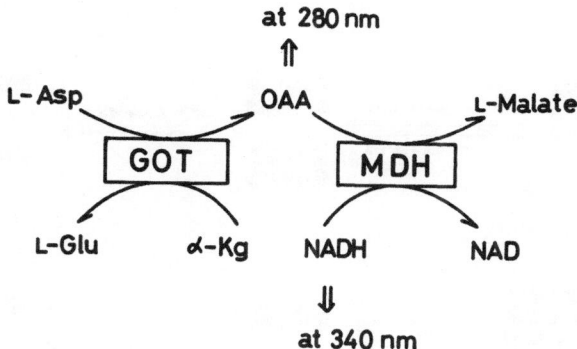

Fig. 9:   Assay of L-aspartate using immobilized GOT or GOT-
          MDH coupled system.

(V)   STUDIES ON THE ACTIVITIES OF SUBUNITS OF ASPARTATE
      4-DECARBOXYLASE

        The relationship between the subunit structures and
the activities of aspartate 4-decarboxylase was first elu-
cidated by immobilizing the dissociated enzyme molecules
on CNBr-activated Sepharose, thus preventing them from re-
association (20). As shown in Fig. 10, this enzyme consists of 16 subunits in its native state, but is known to
dissociate into an intermediary dissociated state (half
the molecular weight of the native form, i.e., octomer) by
treatment with 1 M guanidine-HCl and into an irreversibly
dissociated state (dimeric form) with 5 M guanidine-HCl,
respectively. Since the intermediary dissociated state
(octomer) easily reassociates into the natural state (hexadecamer) under the assay conditions, the activity of the
reversible intermediary state has not been studied as yet,
although such the state would occur in biological systems
according to changes of circumstances such as concentration
change and so on. Upon immobilization, the relative activity of hexadecameric form because ca. 60% of the soluble
state. The activity of immobilized octomer was first confirmed to be ca. 35% of the soluble native enzyme. Dissociation experiments of immobilized hexadecamer to immobilized octomer by treatment with 1 M guanidine-HCl and hybridi-
zation of immobilized octomer with soluble octomer confirmed
the relation between the subunit structures and the activi-

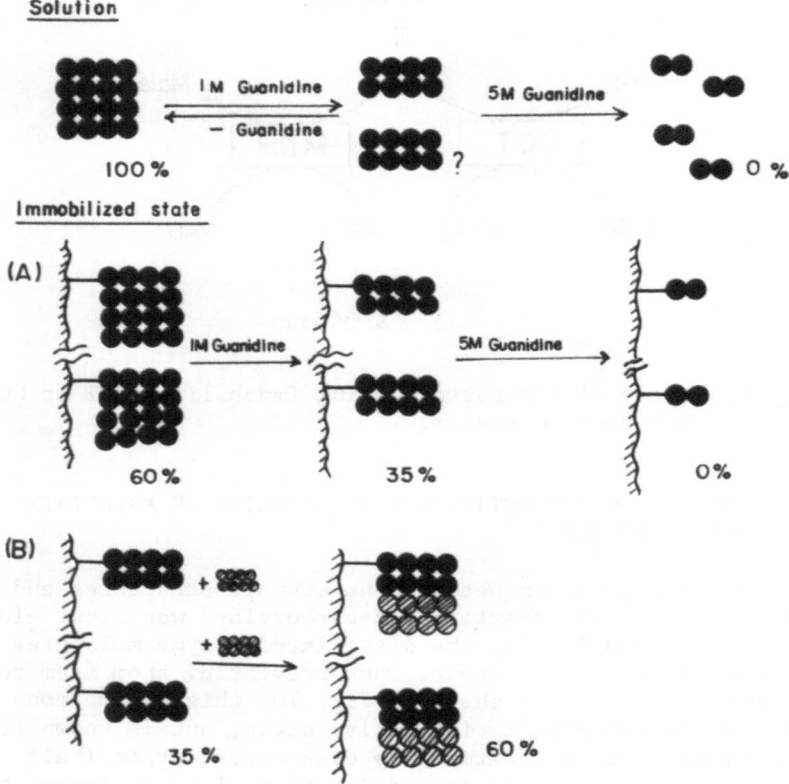

Fig. 10: Changes in subunit structures of aspartate 4-
decarboxylase depending upon circumstances and
application of immobilization method to elucida-
tion of activity changes.

ties of aspartate 4-decarboxylase (Fig. 10).

(VI) KINETIC STUDIES ON COENZYME BINDING AND COENZYME DIS-
SOCIATION OF SOME B_6 ENZYMES IMMOBILIZED ON SEPHAROSE

As illustrated in Fig. 4, PLP was gradually liberated
from holo-Trpase or β-tyrosinase immobilized on Sepharose
during repeated batch reactions or continuous flow reactions.
Supplement of PLP to the reaction mixture quickly restored

the initial activity. Using these phenomena, the binding
rate of PLP to apoenzyme and the dissociation rate of the
coenzyme from holoenzyme were very conveniently studied
kinetically by following the enzyme activity. The details
of this work is going to be published elsewhere (21).   Immobilization methods would be very useful for the studies
on interactions between two or among plural biologically
important substances having complementality.

REFERENCES

1. S. Ikeda and S. Fukui, *Biochem. Biophys. Res. Commun.*,
    52, 482-488 (1973).

2. S. Ikeda, H. Hara and S. Fukui, *Biochim. Biophys. Acta*,
    372, 400-406 (1974).

3. E. McEvoy-Bowe, *Arch. Biochem. Biophys.*, *113*, 167-172
    (1966).

4. W.A. Newton, Y. Morino and E.E. Snell, *J. Biol. Chem.*,
    240, 1211-1218 (1965).

5. Y. Morino and E.E. Snell, *J. Biol. Chem.*, *242*, 2800-
    2809 (1967).

6. T. Watanabe and E.E. Snell, *Proc. Natl. Acad. Sci.*
    U.S.A., *69*, 1086-1090 (1972).

7. H. Kumagai, H. Yamada, H. Mastui, H. Ohkishi and K.
    Ogata, *J. Biol. Chem.*, *245*, 1767-1772, 1773-1777 (1970).

8. H. Yamada, H. Kumagai, N. Kashima, H. Torii, H. Enei
    and S. Okumura, *Biochem. Biophys. Res. Commun.*, *46*,
    370-374 (1972).

9. T. Kakimoto, J. Kato, T. Shibatani, N. Nishimura and
    I. Chibata, *J. Biol. Chem.*, *244*, 353-358 (1969).

10. S.S. Tate and A. Meister, *Adv. in Enzymol.*, *35*, 503-
    543 (1971).

11. W.T. Jenkins and I.W. Sizer, *J. Biol. Chem.*, *235*, 620-
    624 (1960).

12. V.I. Ivanov and M. Ya. Karpeisky, Adv. in Enzymol., 32, 21-53 (1969).

13. S. Fukui, S. Ikeda, M. Fujimura, H. Yamada and H. Kumagai Eur. J. Biochem., in press (1975).

14. S. Fukui, S. Ikeda, M. Fujimura, H. Yamada and H. Kumagai, Eur. J. Appl. Microbiol., in press (1975).

15. H. Enei, H. Nakazawa, H. Matsui, S. Okumura and H. Yamada, FEBS Lett., 21, 39-41 (1972).

16. H. Nakazawa, H. Enei, S. Okumura, H. Yoshida and H. Yamada, FEBS Lett., 25, 43-45 (1972).

17. S. Fukui and S. Ikeda, Process Biochem., in press (1975).

18. S. Ikeda and S. Fukui, FEBS Lett., 41, 216-218 (1974).

19. S. Ikeda and S. Fukui, FEBS Lett., 46, 295-298 (1974).

20. S. Ikeda and S. Fukui, Eur. J. Biochem., 46, 553-558 (1974).

21. S. Ikeda, Y. Sumi and S. Fukui, Biochemistry, in press (1975).

ENZYME ENGINEERING CASE STUDY: IMMOBILIZED LACTASE

James R. Ford and Wayne H. Pitcher, Jr.

Research & Development Laboratory
Corning Glass Works
Corning, New York 14830

The hydrolysis of the lactose in whey to glucose and galactose is becoming an attractive commercial process. As the cost of sweeteners from cane and corn continues its dramatic climb (having almost doubled within the past year), many segments of the food industry are seeking alternative sources of their sugar supply. The use of the glucose isomerase process, where glucose is converted to its much sweeter isomer, fructose, is an example of the food industry's efforts to "stretch" the supply of sweetening sugars.

Acid whey (the solids being $\sim$ 75% lactose), a by-product of cottage cheese manufacture, is of low value and can present a waste disposal problem to cheese producers. On the other hand, the dairy industry has been particularly hard hit by the cost increases of sweeteners, which must be purchased in large quantities for use in ice cream, yogurt, egg nog and other dairy products. Relief from this price squeeze may be available within the dairy industry itself. Converting the lactose present in whey to glucose and galactose increases the sweetness and solubility of whey solids, thus allowing potential use of whey and modified whey as a sweetener substitute in many dairy products.

The conversion of lactose to glucose and galactose is promoted by the enzyme lactase. At the current high lactase enzyme prices, the most promising way of achieving economic feasibility is by using immobilized enzymes. Attaching enzymes to insoluble carriers makes possible their repeated or continuous use over extended periods of time.

17

The studies discussed here were made to determine the feasibility of using immobilized lactase for the commercial hydrolysis of the lactose in cheese whey.

BACKGROUND

Previous studies by Weetall et al (1,2), Olson and Stanley (3), Stanley and Palter (4), Charles et al (5), and Pitcher (6) have demonstrated the general feasibility of lactase immobilization and the use of a lactase system for hydrolysis.

Crude cost estimates (5,6) have also been made. However, sufficient data, including long term stability, reactor sanitization, and additional processing costs, have not been presented.

Weetall et al (1) have discussed the kinetics of lactose hydrolysis, giving the rate equation as

$$v = \frac{kES}{S + K_M \left[1 + (P/K_i)\right]} \tag{1}$$

where v = reaction rate, k = turnover number, E = amount of enzyme, $kE = V_{max}$ (the maximum rate achieved where $S \gg K_m$), S = substrate (lactose) concentration, K_m = Michaelis constant, P = product concentration, and K_i = inhibition constant.

The integrated form of equation (1) for a batch or plug flow reactor, assuming no product present initially, is

$$\frac{Et}{V} \text{ or } \frac{E}{F} = \frac{1}{k}\left\{\left(\frac{K_i - K_m}{K_i}\right)(S_o - S) + \left(\frac{K_m S_o}{K_i} + K_m\right) \ln (S_o/S)\right\} \tag{2}$$

where V = volume of substrate solution, t = elapsed time, S_o = initial or feed substrate concentration, S = substrate concentration at time t or in the reactor effluent, and F = volumetric flow rate of substrate.

Weetall et al (2) reported experimental values of 0.05 M and 0.0039 M respectively for K_m and K_i for soluble Wallerstein lactase.

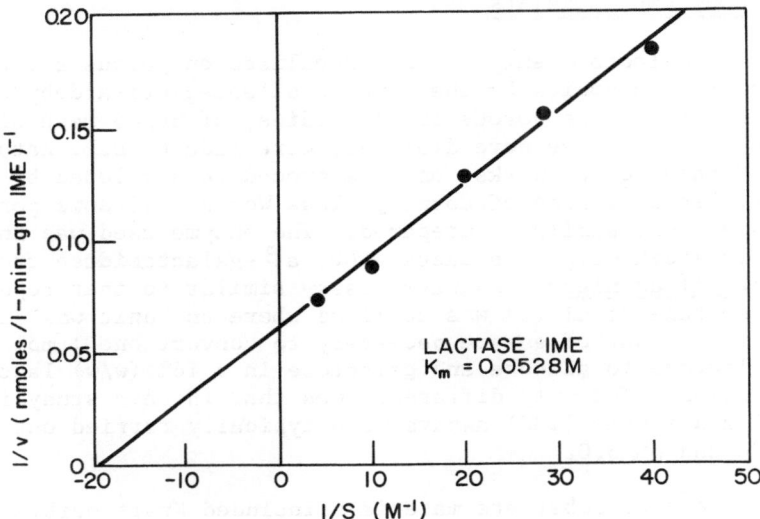

Fig.1: Lineweaver-Burk-type plot of kinetic data for lactase immobilized on SiO_2 with lactose as substrate.

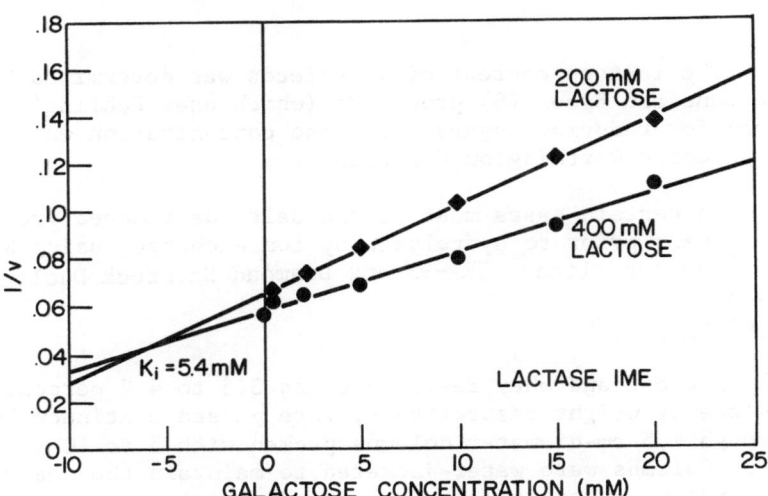

Fig.2: Dixon plot of kinetic data for lactase immobilized on SiO_2 with lactose as substrate and galactose as inhibitor.

METHODS AND MATERIALS

The lactase enzyme was immobilized on porous silica and titania bodies by the aqueous silane-glutaraldehyde method (7). The porous silica bodies, of 30/45 mesh size and 370 Å average pore diameter, were made by D.L. Eaton of Corning Glass Works, using a procedure developed by R.A. Messing, also of Corning Glass Works. Titania porous bodies were similarly prepared. The enzyme used was the Wallerstein Company's Lactase-LP, a β-galactosidase from Aspergillus niger. An enzyme assay similar to that reported by Weetall et al (1) was utilized where one unit was defined as the amount of enzyme necessary to convert one μ mole/min. of lactose to glucose and galactose in a 16% (w/w) lactose solution. The only difference was that in this study immobilized enzyme (IME) assays were typically carried out at 40°C and pH 3.0.

Feed or substrate materials included Kraft edible lactose, acid whey supplied by Crowley Food, Inc. and deproteinized acid whey (ultrafiltrate permeate), also supplied by Crowley. The edible lactose feed solutions contained Zepharin Chloride® .

The lactose content of whey feeds was determined by the Lane and Eynon (8) procedure (which uses Fehling's solution) for reducing sugars. Glucose concentration was measured using Worthington Glucostat .

In certain cases most of the salt was removed from the whey feeds prior to hydrolysis by ion exchange, using Rohm and Haas Amberlite® IRA-93 and Diamond Shamrock Duolite® D-25D.

REACTOR STUDIES

Lactose and whey feeds, 5.0 and 3.3 to 4.0 percent lactose by weight respectively, were passed continuously through 1.5 cm-diameter columns packed with 5 to 10 g of IME. Columns were water-jacketed to maintain the desired temperature. Larger columns, including one 4 inches in diameter, containing about 6 lb. of IME, were also operated. Conditions of operation for specific studies are discussed in the subsequent sections.

IME KINETIC CONSTANTS

Kinetic constants K_m and K_i for the IME were found to be 0.0528M and 0.0054M respectively at 40°C, as shown in Figures 1 and 2.

RATE EQUATION

A theoretical curve relating conversion to normalized residence time (IME activity/flow rate) was calculated from equation (2) using the experimentally determined K_m and K_i values (k = 0.06 when F has units of ml/hr). Experimental data from column operation agreed closely with the theoretical curve as shown in Figure 3. Thus, given conversion and flow rate, equation (2) could be used to obtain activity for monitoring column performance.

pH OPTIMA

The variation in enzyme activity, from initial rate determinations, as a function of pH for soluble and immobilized lactase is shown in Figure 4. Soluble lactase data

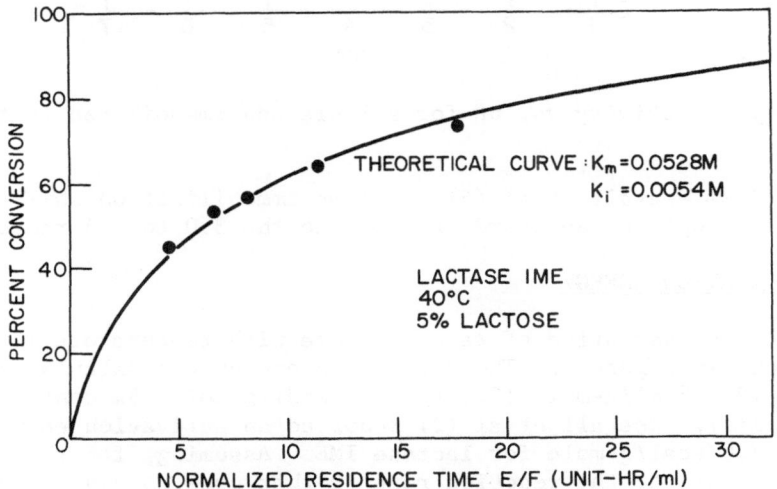

Fig.3: Predicted vs. experimental column performance of lactase IME.

J. FORD & W. PITCHER

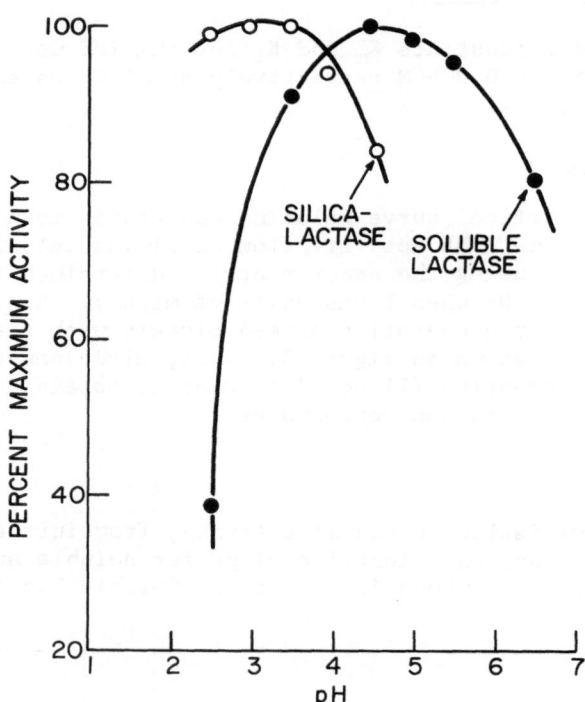

Fig.4: Activity vs. pH for soluble and immobilized lactase.

is from Weetall et al (2). Enzyme immobilization shifted
the pH optimum downward from 4.5 to the 3.0 to 3.5 range.

ACTIVATION ENERGY

The variation of reaction rate with temperature is
shown in Figure 5. The activation energy was calculated to
be 12.0 kcal/g-mole (9.6 to 14.3 kcal/g-mole 95% confidence
limits). Weetall et al (2) reported an activation energy
of 7.8 kcal/g-mole for lactase IME. Assuming, for the sake
of argument, the reaction rate in that case to have been
entirely diffusion controlled, the apparent activation
energy should have been equal to half the intrinsic activation energy plus 1 or 2 kcal/g-mole for diffusivity effects.
The 7.8 kcal/g-mole apparent activation energy is consistent

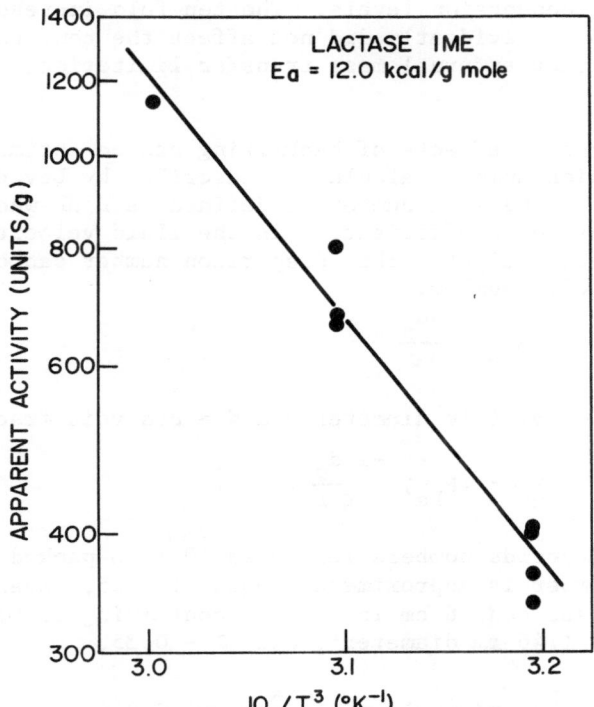

Fig. 5:   An Arrhenius plot of activity data for lactase IME.

with an intrinsic activation energy of 13 or 14 kcal/g-mole
or less. This fact coupled with the straight line plot in
Figure 5 indicates the absence of significant pore diffusion
effects in this study.

MASS TRANSFER STUDIES

Internal mass transfer or pore diffusion limitations
appear to be minimal, at least up to 60°C with the 400 unit/
g (40°C) IME, from the reaction rate versus reciprocal temperature argument invoked in the preceding section.

External mass transfer or film diffusion effects were
evaluated by operating columns containing IME beds ranging
from 6 cm to 71 cm in height. These reactors were operated

with the same normalized residence times and resulted in identical conversion levels. The ten fold increase in linear velocity evidently did not affect the conversion rate, implying that external mass transfer limitations were negligible.

Potential effects of backmixing can be estimated from a dispersion number calculation described by Levenspiel (9). The dispersion number is defined as D/uL where D is the dispersion coefficient, u is the fluid velocity, and L is the bed height. This dispersion number can be related to the Peclet number,

$$N_{Pe} = \frac{u d_p}{D \epsilon}$$

where d_p = particle diameter and ϵ = bed void fraction, by

$$\frac{D}{uL} = (N_{Pe})^{-1} \frac{d_p}{\epsilon L} . \tag{3}$$

At Reynolds numbers less than 10 in a packed bed, the Peclet number is approximately equal to 0.5. Even for a small packed bed, 6 cm in height, containing 30/45 mesh particles (.46 mm diameter), with ϵ - 0.35,

$$\frac{D}{uL} = (\frac{1}{0.5}) \frac{.046}{(35) (6)} = 0.044.$$

From a graph given by Levenspiel for first order kinetics, used as an approximation (actual kinetics should be less sensitive to backmixing), the difference between real and ideal reactor volumes at 80% conversion is less than 8%. At lower conversions or greater bed heights the difference becomes even smaller.

COUPLING EFFICIENCY

Coupling efficiency, defined as

$$\frac{E_{IME}}{E_{initial} - E_{recovered}} \tag{4}$$

where E is enzyme activity, is the percentage of enzyme not recovered from the enzyme solution after attachment,

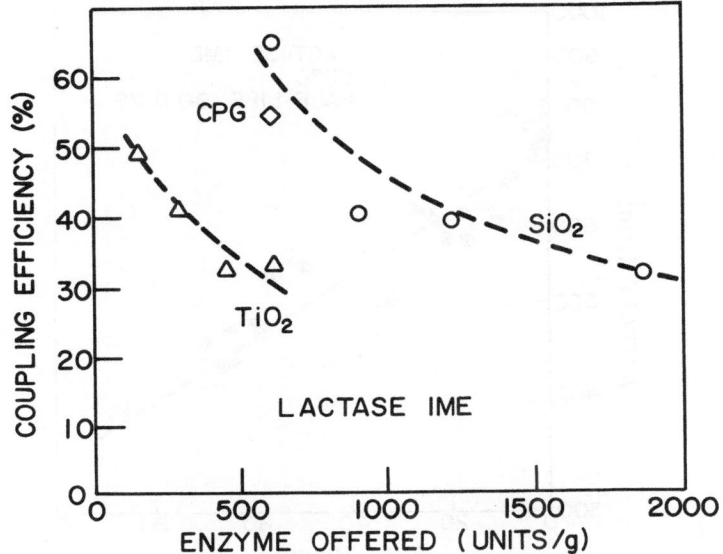

<u>Fig. 6</u>: Coupling efficiency, $E_{IME}/(E_{initial}-E_{recovered})$, as a function of $E_{initial}$.

that is observed as active immobilized enzyme. In Figure 6 the relationship between coupling efficiency and amount of enzyme offered is shown for silica and titania carriers with a reference point for zirconia-coated controlled-pore glass. Much of the difference between the two carriers results from the higher surface area of the silica. In general, the higher the enzyme loading, the lower the coupling efficiency. For cases where the enzyme is relatively expensive, it is important to obtain high coupling efficiency even at the expense of total loading.

<u>HALF-LIFE STUDIES</u>

One of the most important properties of an IME system and most time-consuming to measure is its long-term operational stability. Since this system, like many others, exhibits exponential activity decay, durability is expressed in terms of half-life or the number of days required for the activity to fall to one-half its initial value. Figure 7

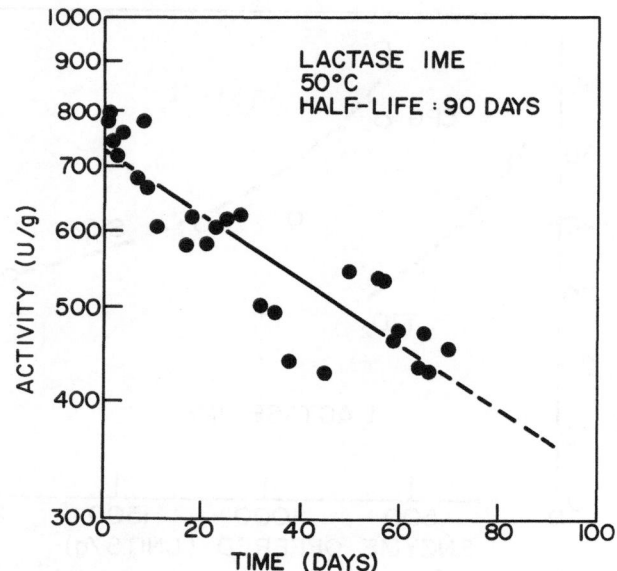

<u>Fig. 7</u>: Activity vs. time for continuous column operation
        of lactase IME.

is an example of this type of data, which falls in a
straight line on a semilogarithmic plot. Half-lives were
calculated from a linear regression of the log of activity
versus time.

Two important variables were found to affect the observed half-life of an IME system: feed composition and
operating temperature. From Table I, it can be seen that
the higher the feed purity, the longer the half-life. The
normal salt content of the whey feed was detrimental to
enzyme half-life although it affected activity insignifi-
cantly. Removal of 90 to 95% of the salt in deproteinized
acid whey resulted in dramatically improved enzyme stability. From the effect of NaCl in lactose on enzyme stability, it would appear that the ionic strength of the salt
plays an important role in decreasing enzyme life. Some of
the difference between the results for ion-exchanged, deproteinized acid whey and lactose may have been due to physical

TABLE I

EFFECT OF FEED COMPOSITION ON HALF-LIFE

Feed	Average Half-Life at 50°C (Days)
Whole Acid Whey	8
De-Proteinized Acid Whey	10
5% Lactose w/ 0.5% NaCl	13
De-Ionized, De-Proteinized Acid Whey	60
5% Lactose	100

attrition from the daily backflushing in the former case.

The effect of temperature on half-life is shown in Figure 8, with a deactivation energy of 40.6 kcal/g mole.

COLUMN BACKFLUSHING

In order to prevent microbial growth in the IME beds where whey feeds were used, the columns were backflushed (fluidized) with distilled water brought to pH 4.0 with acetic acid. Columns were backflushed once daily for about one-half hour. No visible growths have appeared in columns operated in this manner for over a month.

REACTOR SCALE-UP

A four-inch-diameter column reactor containing a 28-inch-deep bed of IME was operated at 38°C. Activity was calculated at 380 units/g (at 38°C) equivalent to 400 units/g at 40°C, identical to the activity normally observed for laboratory scale columns.

Pressure drop, one remaining concern for additional scale-up, can be estimated from Leva's correlation as given by Perry (10).

$$\Delta P = \frac{2 f_m G^2 L (1-\varepsilon)^{3-n}}{\rho d_p g_c \phi_s^{3-n} \varepsilon^3} \tag{5}$$

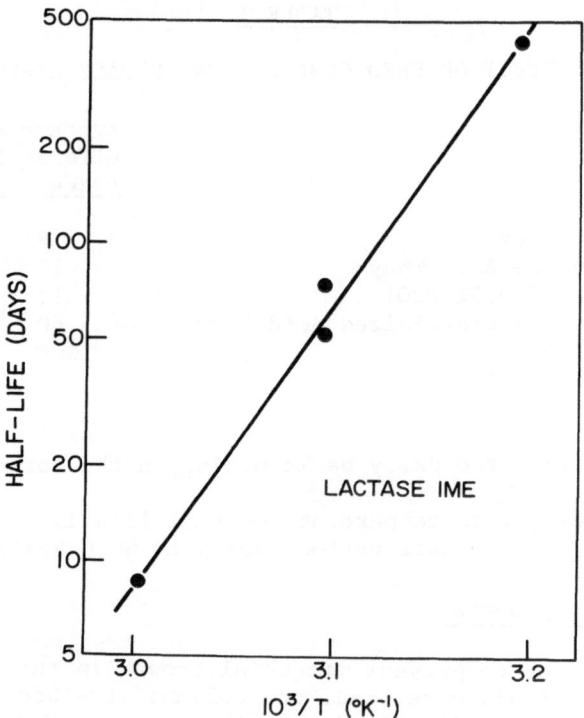

<u>Fig. 8</u>: Half-life vs. receiprocal temperature for lactase
           IME.

where ΔP = pressure (lb.-force/ft^2); L = bed depth (ft.);
G = fluid superficial mass velocity (lb./sec.-ft^2); f_m =
$100/N'_{Re}$ = 100 μ /d$_p$G (for N_{Re} <10), ϵ = void fraction;
n = 1 (for N_{Re} <10), d$_p$ = particle diameter (ft.); g$_c$ =
32.17 (1b-ft/1b force-sec^2); ϕ_s = shape factor (0.95 for
spherical sand), μ = viscosity, ρ = fluid density (1b/ft^3),
and N_{Re} = Reynolds number.

        For the case of a ten-foot-deep IME bed operating at
35°C & 50% hydrolysis a pressure drop of about 75 psi was
calculated. For a six-foot-high column under identical
conditions, a pressure drop of about 27 psi is predicted.

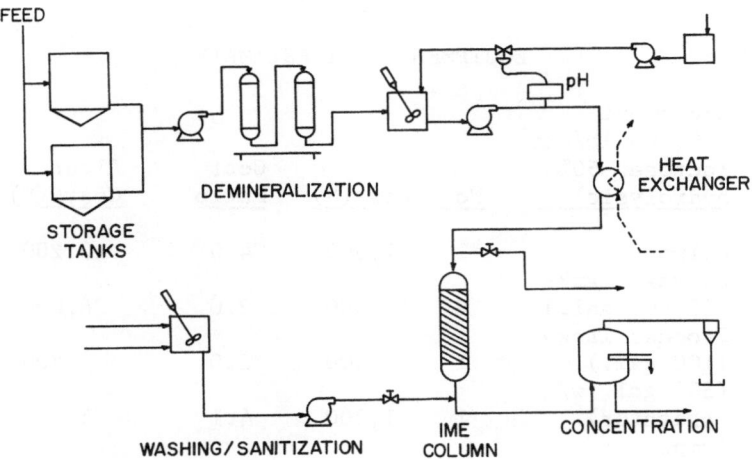

Fig. 9:   Process flow sheet for the recovery and hydrolysis
of lactose from deproteinized acid whey.

PRELIMINARY SYSTEM DESIGN

In order to make a preliminary cost estimate for the
hydrolysis of deproteinized acid whey, a plant design as
shown in Figure 9 was developed. This flow sheet includes
ion exchange and concentration equipment, which will be
discussed independently. The sizes for the hydrolysis system were based on use of lactase IME with an apparent activity of 300 units/g at 35°C. An example of equipment cost
estimates is given in Table II.

The projected operating conditions were set somewhat
arbitrarily with reactor temperature, initially at 35°C,
being raised as necessary to maintain the initial conversion level until 50°C was reached. From a half-life of
62 days at 50°C determined experimentally for deproteinized,
de-ashed acid whey and a deactivation energy of 40.6 kcal/g-
mole, the time required for the 35 to 50°C cycle was calculated at 559 days. The number of lbs of lactose processed
per lb of IME was then calculated from the cycle time and
reactor size.

TABLE II

EQUIPMENT COST ESTIMATE

Equipment (10,000 lb/day lactose, 50% hydrolysis)	No.	Cost($)	Cost Ratio	Plant Cost ($)
Column	1	4,300	4.0	17,200
Storage Tanks (12,500 gal.)	2	18,000	2.0	36,000
Process Tanks (100 gal.)	1	400	2.0	800
(300 gal. w/ agitator)	1	1,300	4.1	5,330
Pumps				
Centrifugal (20 gpm)	3	2,400	8.0	19,200
Metering	1	600	7.0	4,200
Heat Exchanger	1	1,500	4.8	7,200
Instruments		4,000	4.0	16,000
				105,930
Contingency (10%)				10,590
Total..				116,520

Processing costs included labor and supplied cost as shown in Table III, capital or equipment costs taken at 20% annually (included depreciation, maintenance, taxes, etc.) and IME cost. Total costs are reflected in Figures 10, 11 and 12. This total cost does not include the cost of de-ashing or concentration and assumes the cost of deproteinized whey to be zero.

Processing costs appear to be 1 to 4 cent per pound lactose, depending on plant size, IME cost, and percent hydrolysis. Figure 10 shows the dependence of processing cost on plant size at various hydrolysis levels. The effect of IME cost on processing cost is shown in Figure 11. Higher hydrolysis levels obviously are more sensitive to IME costs. Figure 12 shows processing cost as a function of plant size and IME cost. This type of plot can be use-

TABLE III

OPERATING COST ESTIMATE

Labor	(man-hr/day)		
Backflushing	3		
Monitoring	3		
Laboratory	2		
	8 @ $4.50/hr =		$ 36.00
Supervisor	3 @ 6.00/hr =		18.00
			$ 54.00
Overhead & Fringes			54.00
Supplies (Acid, etc.)			20.00
Cooling Costs			3.00
			$131.00

Cost per lb. lactose (10,000 lb lactose/day) 1.3¢

ful for determining what IME cost must be achieved to meet given capacity and cost objectives.

Reducing the salt level by ion exchange or electrodialysis results in projected costs of 2-5 cents per pound, including capital costs comparable to those for the hydrolysis system. Electrodialysis appears economically slightly more favorable at this point, perhaps costing in the vicinity of three cents per pound of lactose for 90 percent salt removal. Similar additional costs will be encountered for concentrating the product from 5 percent solids to the 50 percent or higher solids level necessary for sweetener substitution.

Overall costs of 8 to 10 cents per pound, including royalties, begin to look attractive at the current 15-cent-per-pound (dry basis) price for corn syrups (11).

ACKNOWLEDGEMENTS

The authors gratefully acknowledge the technical assistance of Mr. R.E. Lindner and Mrs. L.F. Bialousz of Corning Glass Works.

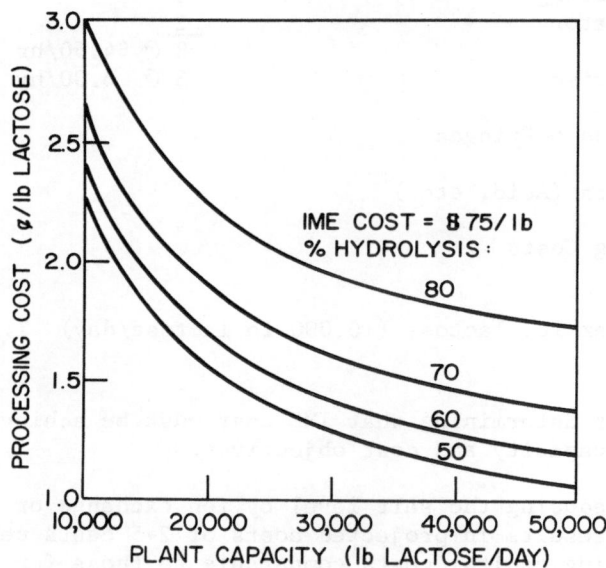

Fig. 10: Processing costs vs. plant capacity for the hydrolysis of lactose from deproteinized, deminer-
aliced acid whey.

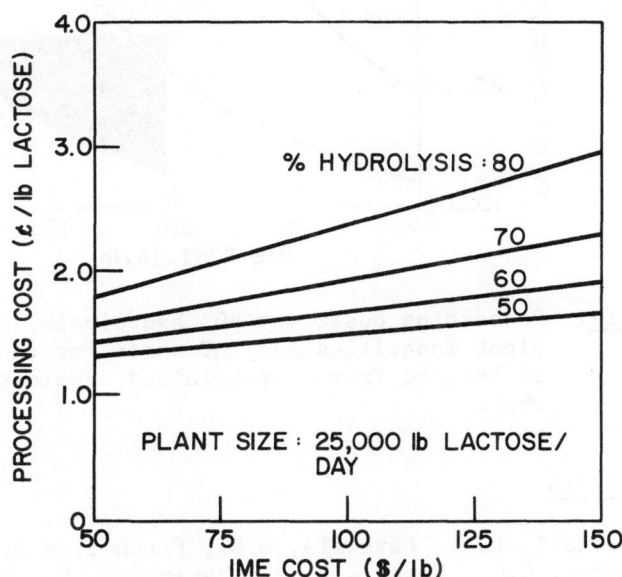

<u>Fig. 11</u>: Processing costs vs. IME cost for the hydrolysis
of lactose from deproteinized, demineralized acid
whey.

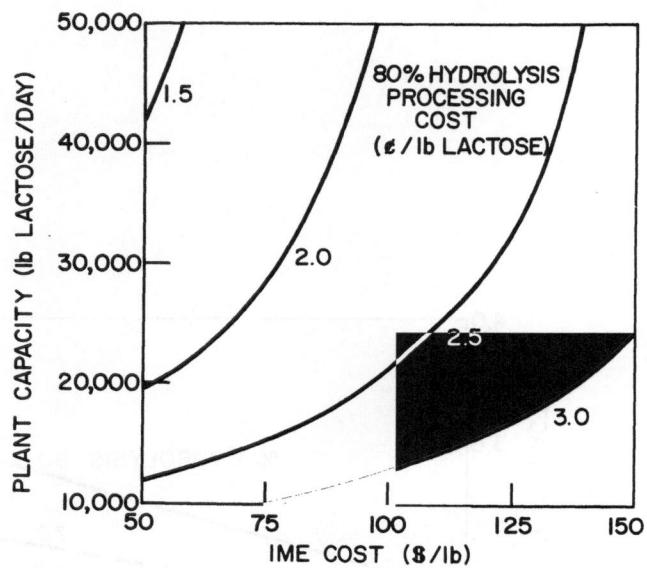

Fig. 12: Processing costs (at 80% hydrolysis) for various
         plant capacities and IME costs for the hydrolysis
         of lactose from deproteinized, demineralized acid
         whey.

REFERENCES

1. Weetall, H.H., Havewala, N.B., Pitcher, W.H., Jr.,
    Detar, C.C., Vann, W.P. and Yaverbaum, S., *Biotechnol.
    Bioeng.* 16 , 295 (1974).

2. Weetall, H.H., Havewala, N.B., Pitcher, W.H., Jr.,
    Detar, C.C., Vann, W.P. and Yaverbaum, S., *Biotechnol.
    Bioeng.* 16, 689 (1974).

3. Olson, A.C. and Stanley, W.L., *J. Agr. Food Chem.* 21,
    440 (1973).

4. Stanley, W.L. and Palter, R., *Biotechnol. Bioeng.* 15,
    597 (1973).

5. Charles, M., Coughlin, R.W., Allen, B.R., Paruchuri,
    E.K. and Hasselberger, F.X., "Increasing Economic Value
    of Whey Wastewaters Using Immobilized Lactase", paper
    17b, presented at AIChE 66th Annual Meeting, Philadel-
    phia (1973).

6. Pitcher, W.H., Jr., Dairy Scope, 18 (March, 1974).

7. Weetall, H.H. and Havewala, N.B. in "Enzyme Engineer-
    ing", L.B. Wingard, Jr., ed. 249, John Wiley, New
    York (1972).

8. Standard Analytical Methods of the Member Companies of
    the Corn Refiners Association, First Revision, 5-27-68,
    E-26.

9. Levenspiel, O., "Chemical Reaction Engineering", John
    Wiley, New York (1962).

10. Perry, J.H., "Chemical Engineers' Handbook", 4th ed.,
    McGraw-Hill, New York (1963).

11. Chemical Marketing Reporter, September 16, 1974.

ALTERATION OF ENZYMATIC CHARACTERISTICS OF BACTERIAL ALPHA-AMYLASE BY ACYLATION

Hirosuke Okada and Itaru Urabe

Dept. of Ferment. Technol., Osaka University
Yamadakami, Suita-shi, Osaka 565,
Japan

INTRODUCTION

Immobilization of an enzyme by chemical bonding to a carrier molecule usually results in some alteration of its enzymatic characteristics. At present the alteration is explained in terms of a new microenvironment generated around the immobilized enzyme that is affected by the physicochemical nature of the carrier molecule (1,2). According to our hypothesis, an enzyme immobilized by conjugate bonding is a kind of modified enzyme; if the modifying group is large enough and insoluble the resulting enzyme is immobilized, but if the group is relatively small and soluble the enzyme obtained is said to be modified. If we choose a suitable series of modified enzymes, they will cover the range of enzymatic characteristics between the natural and the immobilized enzyme.

Based on the above hypothesis, a model experiment was carried out using bacterial α-amylase of the liquefying type and fatty acids of various carbon chain lengths. We compared the characteristics of the acylated α-amylase in relation to the length of the acyl group and the average number of acylation. As the acyl group, acetyl, butyryl, caprylyl and palmityl groups, C_2, C_4, C_8 and C_{16} were used (3).

MATERIALS AND METHODS

Crystalline bacterial α-amylase of the liquefying type was used in the present work. The acylation method employed

37

TABLE 1

Sample	Km(%)	Sample	Km(%)	Sample	(Km(%)	Sample	Km(%)
A- 0	0.05	B- 0	0.05	C- 0	0.05	P- 0	0.05
A- 3	0.05	B- 3	0.05	C- 2	0.05	P- 2.5	0.04
A- 5	0.05	B- 5	0.06	C- 4.5	0.05	P- 4.5	0.04
A- 8.5	0.04	B- 8	0.05	C- 9.5	0.05	P- 7.5	0.04
A-10.5	0.04	B-11	0.08	C-14.5	0.06	P-10.5	0.04
A-14	0.08	B-12.5	0.08	C-17	0.06	P-13	0.05
A-16	0.07					P-16.5	0.05

in this experiment was as follows: α-Amylase at a final concentration of 0.04 mM was allowed to react with the p-nitrophenyl esters of fatty acids at various concentrations in 70% dioxane solution. The enzyme derivatives formed were precipitated by the addition of soluble starch solution in 70% dioxane and dissolved in 0.2 M borate buffer (pH 8.5) containing 0.1 mM calcium chloride. Both acylate α-amylase and soluble starch are soluble in 70% dioxane but the enzyme-substrate complex is insoluble.

Higher concentrations of the ester and longer reaction times resulted in increases in the acylation number.

The number of free amino groups was determined using 2,4,6-trinitrobenzenesulfonic acid according to the procedure of Habeeb (4).

RESULTS

Michaelis Parameters

To determine the effect of substrate concentration on the enzyme activity, 23 acylated α-amylase samples were investigated and the results are shown in Table 1. The table indicates that the K_m value with respect to soluble starch is not affected by acylation. But V_m values decreased with the increase in the chain length of the acyl groups and with the increase in the acylation number as shown in Fig. 1 where V_m values are expressed as percentages of that of the native enzyme.

The decrease in activity by immobilization is usually

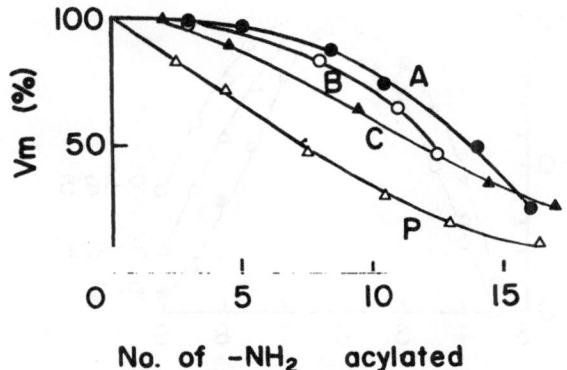

<u>Fig. 1</u>: Effect of acylation of α-amylase on V_m value.

independent of the immobilization technique employed. In
our model case, the decrease in activity could be explained
by amino group modification depending on the size and solubility of the modifying group. The heavily caprylated and
palmitylated enzyme derivatives lost their solubility in
water and formed colloidal suspensions.

pH-Activity Profiles

     The pH-activity curve was shifted to the alkaline side
by palmitylation compared with that of the natural enzyme
as shown in Fig. 2. The degree of the shift by acylation
increased with the increases in acylation number and acyl
carbon chain length. The pH-activity curves of other acylated enzymes also showed the same tendency.

Mode of Amylase Action

     Amylases have been classified into three groups ac-
cording to their mode of action: 1.) single chain mechanism
of glucoamylases, 2.) multiple attack of α-amylase of the
saccharifying type, and 3.) multichain mechanism of α-amylase
of the liquefying type. If the mode of action of bacterial
α-amylase is altered by acylation toward that of multiple

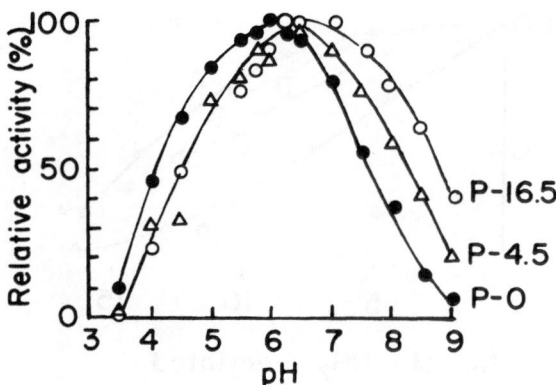

<u>Fig. 2</u>: pH-activity Curve of Palmitylated α-amylase.

attack, the decrease in V_m value could be explained by an
alteration in the mode of action. To test this possibility,
the relationships between the blue value decreases and the
reducing value increase during starch hydrolysis by acetylated and palmitylated enzymes were analyzed. The results
are shown in Figs. 3 and 4. Acetylated α -amylase of acylation numbers up to 16 show almost the same correlation
between the two values. Similar results were obtained using
palmitylated samples, but the heavily palmitylated sample
showed a slight shift toward the multiple attack mechanism.

Heat Stability

     It is often reported that immobilized enzymes are more
heat stable than natural enzymes. The heat resistance of
the acylated enzyme samples was tested at 75°C. Fig. 5
shows the heat inactivation kinetics of the acetylated enzymes. The heat resistance of bacterial α -amylase was
enhanced by acetylation. The acylated enzyme samples used
throughout this study were mixtures of enzyme molecules
having different acylation numbers and sites, and the acylation numbers were specified as mean values. These samples,
when analyzed by acrylamide gel electrophoresis, gave a
single but wider protein band of increased mobility. The

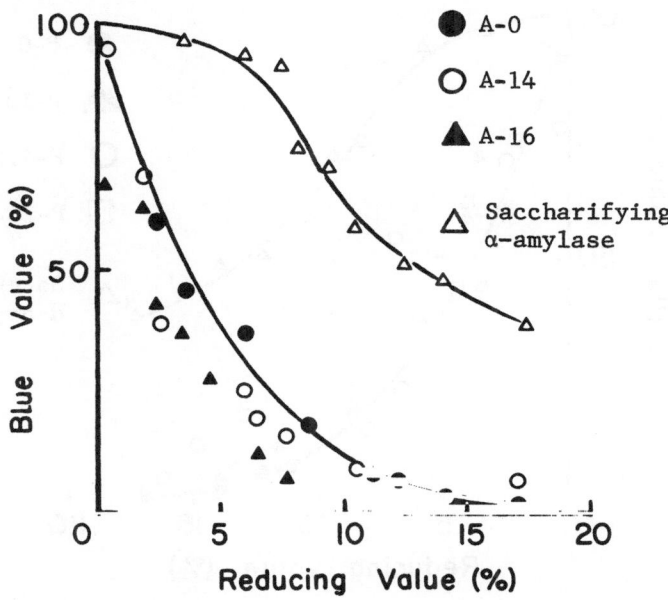

<u>Fig. 3</u>: Relationship between iodine color and reducing
        values during hydrolysis of soluble starch by ace-
        tylated α-amylase.

increase in mobility corresponded to the increase in acylation number. Thus, the acylated enzyme samples were considered to be mixtures of acylated enzymes having acylation
numbers distributed over a narrow range, rather than mixtures of non-acylated and heavily acylated enzymes. This
affects the linearity of the heat inactivation curve as
shown in Fig. 5. From the figure, the apparent heat inactivation rate constants were calculated using the data at
0 and 20 min and plotted against the mean acylation number
(Fig. 6). Acylated α-amylases, especially the acetylated
and butyrylated enzymes, were more heat stable than the
native enzyme.

Thermodynamic Parameters of Heat Inactivation Reaction

        It is generally believed that thermostability of an
enzyme is decided principally by its amino acid sequence

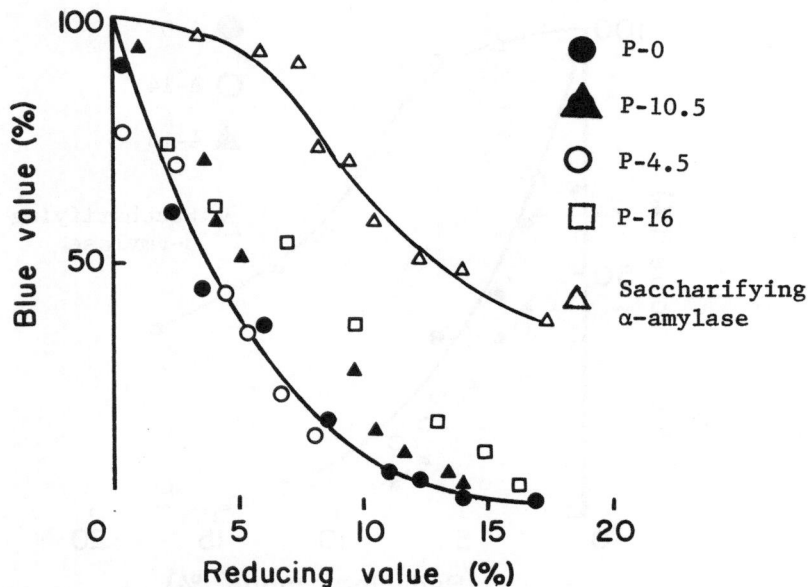

Fig. 4: Relationship between iodine color and reducing
values during hydrolysis of soluble starch by palmitylated α -amylase.

and the specific conformation derived from it. Thus the
heat stability of modified enzymes which have the same pri-
mary structure as some modified amino acid residues and whose
conformations probably differ only slightly from that of the
original is of interest. The effect of acetylation on the
thermostability of α -amylase was investigated at temperatures ranging from 60 to 75°C as shown in Fig. 7. From the
figure, it is evident that the A-11.5 (acetylation number of
11.5) sample was less stable at 60°C than the native enzyme
but more stable at 75°C and had an almost same stability at
67°C. This tendency was more marked when α -amylase of
higher acetylation number was used.

From the data shown in the figure, the apparent heat
inactivation rate constant, k', was estimated and used for
calculation of the activation parameters. Their Arrhenius
plots are shown in Fig. 8. It is of particular interest

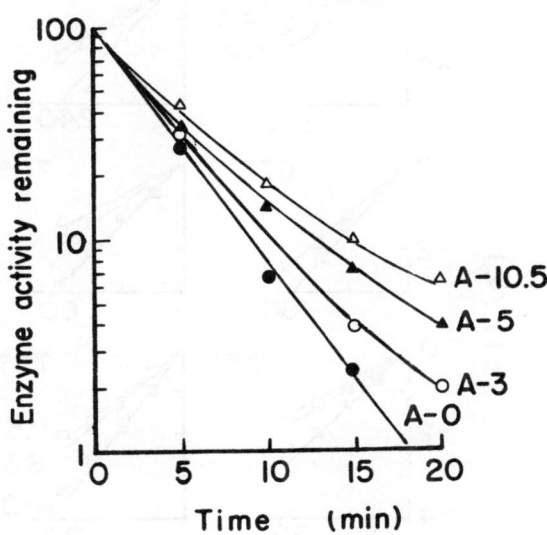

<u>Fig. 5</u>: Heat inactivation curve of acetyl α-amylase.

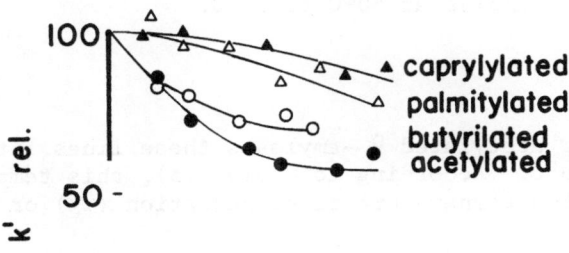

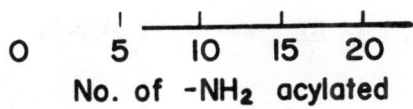

<u>Fig. 6</u>: Effect of acylation on heat stability at 75°C.

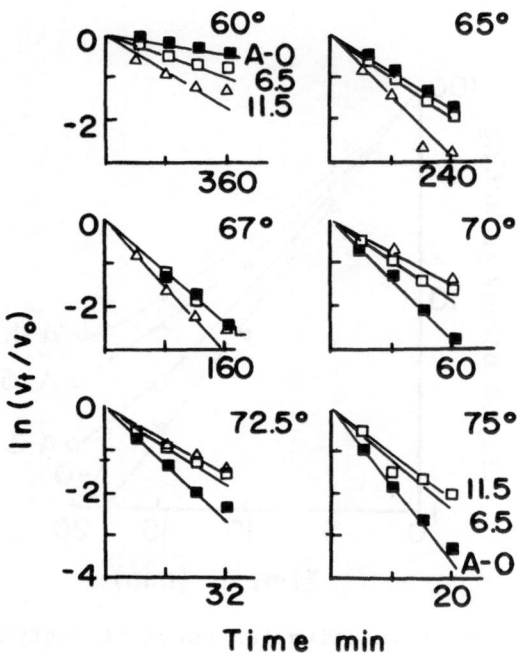

<u>Fig. 7</u>: Heat inactivation curve of acetylated α -amylase
           samples at 60°C to 75°C.

that for acetylated α -amylases these lines intercept at
about 68°C. According to Cremer (5), this temperature is
called the temperature of compensation (T_c) or isokinetic
temperature.

    The activation parameters were determined by applying
the Eyring absolute rate equation:

$$k' = K \frac{KT}{h} e^{-\Delta H*/RT} e^{\Delta S*/R}$$

where K is the transmission coefficient (assumed to be unity)
K is the Bolzmann constant, h is Plank's constant, R is the
gas constant, T is the absolute temperature, and $\Delta H*$ and
$\Delta S*$ respectively are the enthalpy and entropy of activation

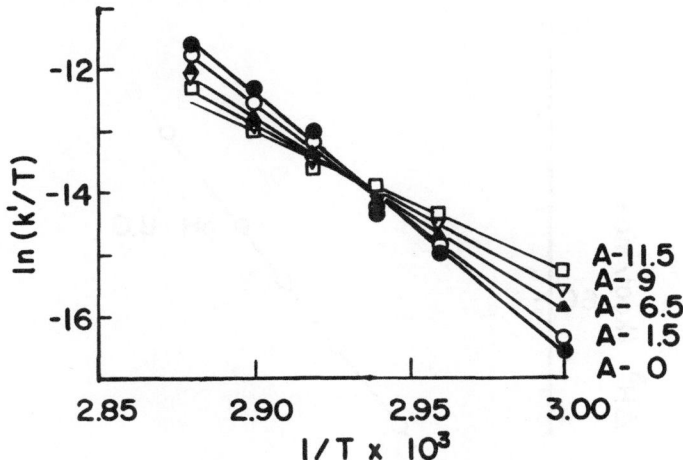

Fig. 8:   Arrhenius plot of heat inactivation of acetylated
          α -amylases.

of the heat denaturation process. $\Delta G*$, Gibbs free energy
of activation, was obtained from the following relationship:

$$\Delta G* = \Delta H* - T \Delta S*.$$

The calculated data showed that the values of $\Delta H*$ and
$\Delta S*$ decreased with the increase in the acetylation number
of the α-amylase, and that the $\Delta G*$ value was constant for
all five samples studied.

Compensation Effect

Another aspect of the compensation effect is the linear
relationship between $\Delta H*$ and $\Delta S*$ (6) as shown in Fig. 9,
in which the slope gives the compensation temperature. The
T_c value obtained was 70°C. In general, the compensation
effect is considered to be a phenomenon in which both $\Delta H*$
and $\Delta S*$, or ΔH and ΔS, of a series of chemical reactions
show a parallel increase in accordance with changes in the
chemical structure of the reactant or in the reaction con-

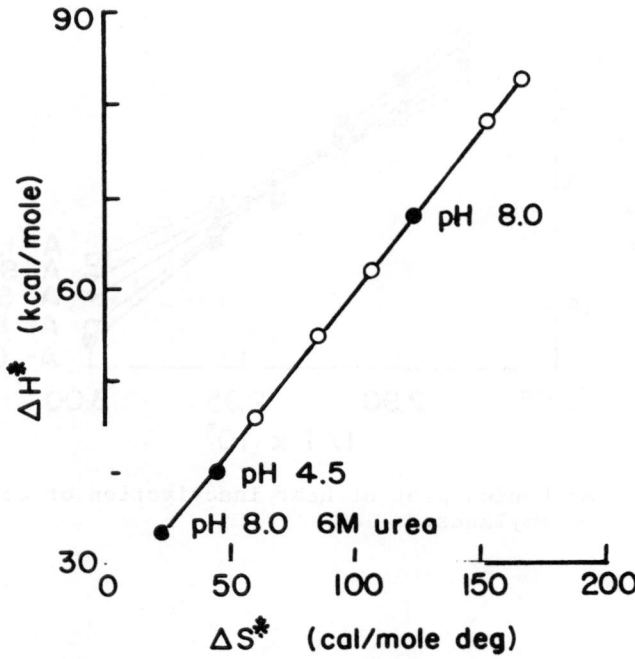

<u>Fig. 9</u>:   Correlation between $\Delta H*$ and $\Delta S*$ of inactivation
         kinetics of acetylated and native α-amylases.

ditions. Such an effect indicates that the reactions in
the series proceed by the same reaction mechanism, or pass
through a common transient state. Lumry et al (7) have
pointed out that a reaction having a compensation temperature between -23 to 42°C is one in which water molecules
are involved. Examples cited by Lumry are ionization and
solubilization of small molecules, denaturation of ribonuclease and trypsin, and the enzyme reaction of α-chymo-
trypsin. Barnes et al (8) also found a compensation effect
in the heat inactivation of viruses with or without a pro-
tecting inorganic salt.

        Unfortunately, the physico-chemical meaning of the
compensation effect is not clear, but it seems reasonable

TABLE 2

T_C Values of Various Protein Denaturation

	pH	$(NH_4)_2SO_4$ M	Urea M	Alcohol (vol.%)	T_C
Hemoglobin	4.08-8.0	0-1.52	0	0	77
Hemoglobin	6.0 -7.0	0	0	0-30	-9
Egg albumin	1.02-9.8	0 or 1	0 or 1	0	79
Invertase	3.0-5.7	0	0	0	55
Pancreatic ribonuclease	7.3	0	0-8.0	0	40
B. subtilis α -amylase	4.5-8.5	0	0-6	0	70

to conclude that the mechanism of heat inactivation of α - amylase remained unchanged by acetylation.

Yamanaka et al (9) have studied the thermal inactivation kinetics of α -amylase produced by B. subtilis under various conditions (at pH 8.0, pH 4.5 and pH 8.0 in the presence of 6M urea). ΔH* and ΔS* values calculated from their data were also plotted on Fig. 9. These three plots fell onto the same line. Thus it can be seen that the change in heat resistance of α -amylase by acetylation, and probably by immobilization, is the same phenomenon as the change in thermal stability of an enzyme under different experimental conditions, and can be explained by the compensation effect. The decrease in ΔH* and ΔS* by acetylation indicates that the molecular conformation of α -amylase is shifted toward a random activated state.

The compensation effect in the thermal denaturation of proteins and enzymes was generally observed in the data presented by other authors (9,10,11). From available experimental data, isokinetic temperatures were calculated and summarized in Table 2. For example, in the case of pancreatic ribonuclease (10), data include denaturation and renaturation at pH 7.3 in urea solutions of various concentrations, but all observed data can be explained by a single isokinetic line of T_C 40°C. This might suggest that denaturation and renaturation involve the same reaction mechanism in opposite directions.

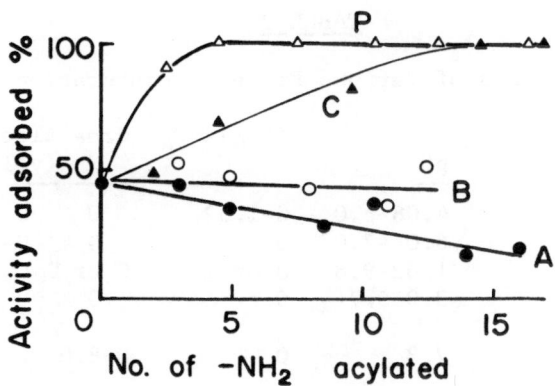

<u>Fig. 10</u>:   Effect of acylation number on adsorption on Milli-
            pore filter HA 45.

Deviation from an isokinetic line indicates a different
mechanism for the reaction. In the case of hemoglobin denaturation (11), the series of heat denaturation reactions
can be explained by an isokinetic line of T_C 77ºC, and
another series of denaturation reactions in the presence
of ethanol had an isokinetic line of T_C-9ºC. So in this
case, we can judge that heat denaturation and alcohol denaturation proceed by different mechanisms.

The compensation effect could be extrapolated to in-
solubilized enzymes, which in many instances were reported
to be more heat resistant than the native enzyme. In many
reports, heat stability was measured at a constant temperature, but from the general standpoint described above, the
existence of the compensation effect is highly probable.

Immobilization of Acylated α-Amylase

Acylated α-amylase can be fixed on a Millipore filter
and can be used for continuous reaction. Merely by passing
the enzyme solution through a Millipore filter, HA 45, a
remarkable amount of acylated α-amylase could be immobilized
on the filter. The amounts of enzyme immobilized on the membrane were estimated from the differences in protein concen-

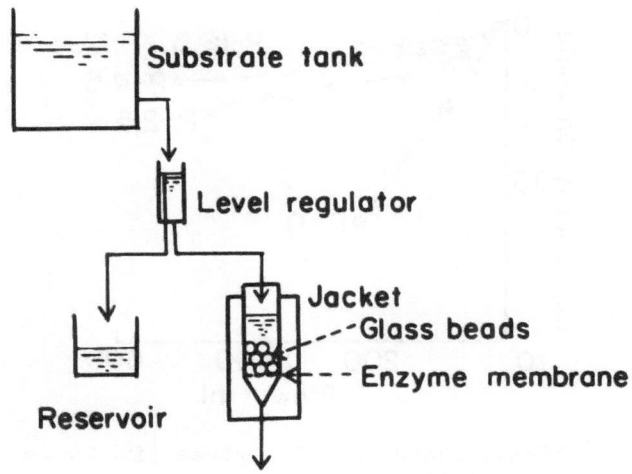

<u>Fig. 11</u>:   Flow sheet of continuous reactor.

tration or enzyme activity of the solutions before and after
filtration, and were plotted against the acylation number,
as shown in Fig. 10. Even for the native enzyme, 45 per
cent of the activity was adsorbed onto the filter, but, as
will be shown later, the activity quickly disappeared from
the membrane on passing substrate solution. Palmitylated
and heavily caprylylated enzymes were fixed quantitatively.
The affinity of the acetylated α-amylase for the membrane
decreased in accordance with the increase in the acetylation number. These results indicate that the binding force
of the native enzyme is ionic, whereas that of the palmitylated enzyme is hydrophobic. The enzyme activity was reduced by 40 to 50% on immobilization.

Acylated α-amylase, fixed on the membrane surface,
can be used for continuous hydrolysis of soluble starch
using the apparatus shown in Fig. 11. The substrate solution from the storage vessel was introduced into the reactor column after passing through a level regulator. In the
column, glass beads were packed to a height of 1 cm to avoid

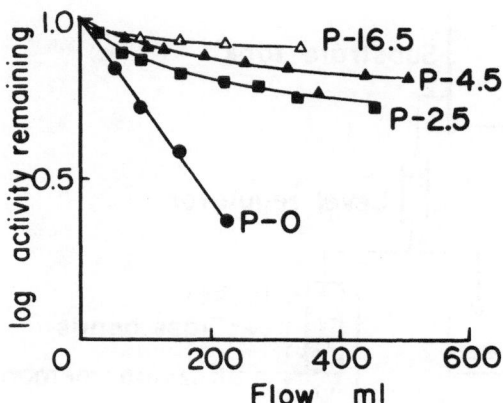

<u>Fig. 12</u>: Activity change of α-amylase fixed on a Millipore
filter HA 45 during continuous hydrolysis of solu-
ble starch.

counter flow of substrate solution and at the bottom of the
column the enzyme membrane was placed.

During continuous reaction, the native enzyme easily
leaked from the membrane by mono-molecular reaction, but
the heavily palmitylated enzyme samples lost their activities only during the early stage of continuous flow and
later the activities remained constant as shown in Fig. 12.

From the data obtained above, the yield of α-amylase
activity from the native enzyme which could be stably fixed
on a millipore filter was calculated to be 28%.

The technique of immobilizing an acylated enzyme on a
membrane appears unpractical at this stage, but after development of an ultrafiltration membrane which can separate
substrate and coenzyme molecules by filtration, this technique would be useful for a continuous enzyme reaction which
requires coenzyme.

REFERENCES

1. Hornly, W.E.; Lilly, M.D.; Crook, E.M.; Biochem. J. <u>107</u>,

669 (1968).

2. Filippusson, H.; Hornby, W.E.; Biochem. J. 120, 215
    (1970).

3. Urabe, I.; Okada, H.; Ferment. Technol. Today (Proc.
    IV IFS) 367 (1972).

4. Habeeb, A.F.S.A.; Anal. Biochem. 14, 328 (1966).

5. Cremer, E.; Adv. Catalysis 7, 75 (1955).

6. Leffler, J.E.; J. Org. Chem. 20, 1202 (1955).

7. Lumry, R.; Rajender, S.; Biopolymers, 9, 1125 (1970).

8. Barnes, R.; Vogel, H.; Gordon, I; Proc. Natl. Acad.
    Sci. U.S. 62, 263 (1969).

9. Yamanaka, T.; Higashi, T.; Horio, T.; Okunuke, K.;
    J. Biochem. 44, 637 (1957).

10. Nelson, C.A.; Hummel, J.P.; J. Biol. Chem. 237, 1567
    (1962).

11. Eyring, H.; Stearn, A.E.; Chem. Rev. 24, 253 (1939).

GLUCOSE ISOMERASE IMMOBILIZED ON PHENOL-FORMALDEHYDE RESIN

Y. Yokote, K. Kimura and H. Samejima

Tokyo Research Laboratory
Kyowa Hakko Kogyo Co., Ltd.
Tokyo, Japan

In order to apply an immobilized enzyme in an industrial process, its preparative method especially the support material used for the preparation must be selected from the standpoint of its economy and feasibility in a large scale operation. From this reason, the authors had chosen phenol-formaldehyde resins as promising support materials for immobilization of various enzymes and reported at the 2nd Enzyme Engineering Conference in 1973 (1).

On the other hand, glucose isomerase (hereafter abbreviated as GIS) is getting world-wide attention because of its capability in conversion of glucose to fructose, and also the increasing demand of pure fructose and high fructose syrup for food and pharmaceutical uses. Therefore, several kinds of immobilized GIS were already reported, e.g., GIS adsorbed on DEAE-Sephadex (2), GIS entrapped with polyacrylamide (3-4) and cellulose triacetate (5), GIS covalently bound on porous glass (6), GIS stabilized in heat-treated microbial cells (7), and so forth.

The authors prepared two types of immobilized GIS using a phenolformaldehyde resin, i.e., Duolite A7, as a support material. One type of immobilized GIS is an adsorption type (hereafter, abbreviated as Duolite A7-ADS-GIS) and the other is a covalently binding type using triazinyl chloride as a coupling agent (hereafter, abbreviated as Duolite A7-CVB-GIS).

The preparative methods, properties, stabilities and continuous operations of such immobilized GIS were studied and reported in the present paper.

53

MATERIALS AND METHODS

Enzyme used. Frozen mycelia of GIS producing fungus, i.e.,
Streptomyces phaeochromagenes (purchased from the Nagase
Sangyo Co., Ltd. Japan) were homogenized by a Manton-Gaulin
homogenizer, and the supernatant was treated successively
by ammonium sulfate precipitation, dialysis, and heat treatment. The partially purified GIS thus obtained was then
used as the native enzyme throughout the following experiments.

Preparation of Duolite A7-ADS-GIS. A phenol-formaldehyde
resin, i.e., Duolite A7 (a weak anion exchange resin produced by the Diamond Shamrock Chemical Co., U.S.A.) was
bufferized with 0.05 M sodium acetate-hydrochloric acid
buffer (pH 7.7). The bufferized resin was contacted with
a GIS solution in a buffer (pH 8.0) containing 0.05 M
$NaHCO_3$, 0.05 M $MgSO_4$, and 0.5 mM $CoCl_2$; and then incubated
for 16 hours at room temperature with continuous stirring.
Thus prepared Duolite A7-ADS-GIS was filtered off and
washed with the same buffer and 5 M NaCl successively.

Preparation of Duolite A7-CVB-GIS. A bifunctional cross-
linking agent, i.e., triazinyl chloride (8), was used for
preparing 2,4-dichloro-a-triazinyl Duolite A7. Thus prepared 2,4-dichloro-s-triazinyl Duolite A7 was washed with
0.1 M phosphate buffer (pH 7.5), and then contacted with a
GIS solution containing 0.05 M $MgSO_4$ and 1 mM $CoCl_2$ and
incubated for 20 hours at 5°C with continuous shaking.
After the reaction, the enzyme resin complex was filtered
off and washed with 0.05 M phosphate buffer (pH 8.0) and
5 M NaCl successively.

Measurement of enzyme activity. Five ml of the immobilized
GIS were packed in a glass column (1 x 8 cm, D/H). A substrate solution which comprised 0.5 M glucose, 0.01 M $MgSO_4$,
0.5 mM $CoCl_2$ and 0.05 M $NaHCO_3$ buffer (pH 8.2) was passed
through the column with a space velocity (S.V.) of 10 at
60°C.

The units of enzyme activity were expressed with the amounts
of fructose produced (μ moles) by one ml of the immobilized
enzyme for one minute. In this case, fructose was assayed
by thiobarbituric acid method.

TABLE 1

ELUTION PATTERNS OF ENZYME PROTEIN FROM THE IMMOBILIZED
GIS USING VARIOUS ELUANTS

Immobilized GIS	Amount of Protein (mg/ml-R)				
	Charged	Eluted by			Not Eluted
	Buffer	5M-NaCl	2N-NaOH		
	(a)	(b)	(c)	(d)	(e)
Duolite A7-ADS-GIS	40.3	27.9	0.6	4.6	6.6
Duolite A7-CVB-GIS	40.3	26.1	0.0	0.8	13.5

RESULTS

Binding of Enzyme Protein and Enzyme Activity on the Resin

Two types of immobilized GIS just after the immobilization reactions were packed in glass columns respectively,
and eluted with 0.05 M phosphate buffer (pH 8.0), 5 M NaCl
and 2 N NaOH successively. Such elution patterns are shown
in Table 1. Nearly half amount of the enzyme protein fixed
on the resin was eluted by 5 M NaCl and 2 N NaOH in the
case of Duolite A7-ADS-GIS. On the contrary, very small
amount of the enzyme protein was eluted by the same eluants
in the case of Duolite A7-CVB-GIS. These observations suggested that Duolite A7-ADS-GIS bound the enzyme protein
mainly by physical adsorption and/or ionic binding, and on
the contrary Duolite A7-CVB-GIS bound the enzyme protein
mainly by covalent binding.

The amounts of enzyme protein bound, enzyme activities
assayed, and ratio of enzyme activity expressed are summarized in Table 2. Covalent binding type, i.e., Duolite
A7-CVB-GIS, shoed better results both in enzyme activity
assayed and in ratio of enzyme activity expressed. In
later experiments, Duolite A7-ADS-GIS were prepared with
using various native enzymes from different purification
stages, and the enzyme activity and ratio of the enzyme
activity expressed were also determined. These results
are shown in Table 3. It is apparent that the grade of
purification does not affect so much for the enzyme activity

TABLE 2

ENZYME PROTEIN AND ACTIVITY BOUND ON THE IMMOBILIZED GIS

Immobilized GIS	Protein Bound (mg) (a)-(b)-(c)	Enzyme Activity Assayed i.u./ml resin	Ratio of Enzyme Activity Expressed %
Duolite A7-ADS-GIS	11.8	21.4	28.0
Duolite A7-CVB-GIS	14.2	34.0	36.6

TABLE 3

ENZYME ACTIVITIES OF DUOLITE A7-ADS-GIS PREPARED WITH
DIFFERENT GRADES OF NATIVE ENZYME

Step	Purification		A7-ADS-GIS	
	Yield %	Specific Activity i.u./mg-p	Activity i.u./ml-R	Ratio of Activity Expressed %
Homogenate	79.5		31.6	35.4
$(NH_4)_2SO_4$ 15% sup	72.0	6.5	14.7	72.2
$(NH_4)_2SO_4$ 50% ppt Dialysate	44.5	9.7	36.0	27.2
Heat Treat sup	50.9	12.4	36.2	30.0
$(NH_4)_2SO_4$ 70% ppt Dialysate	40.0	12.5	35.0	27.4

Charged GIS: 121 i.u./ml-R

and the ratio of enzyme activity expressed in the immobilized GIS preparations. Even when the mycelial homogenate
was directly used as the native enzyme, satisfactory enzyme
activity and retention of activity of the immobilized GIS
preparation were attained. This suggested that selective
adsorption of GIS from the homogenate was conducted by Duolite A7. And, this phenomenon is very favorable for the
industrial preparation of the immobilized GIS.

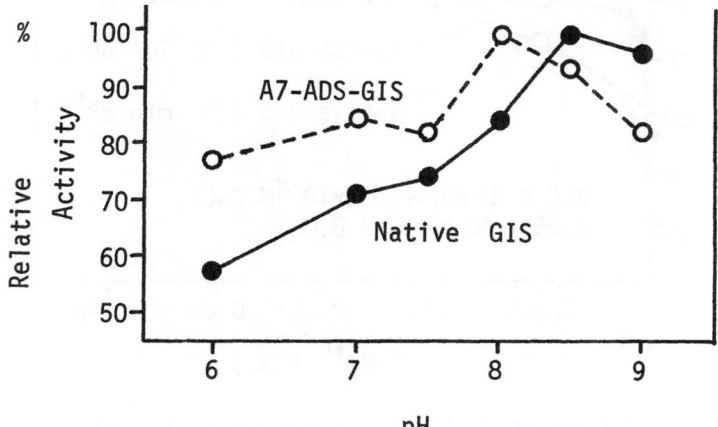

<u>Fig. 1</u>:   Effect of pH on Activity of A7-ADS-GIS

Properties of Immobilized GIS

<u>Optimum pH and temperature for enzyme activity</u>.   As
shown in Figure 1, native GIS has its optimum pH for the
enzyme activity at 8.5.   However, Duolite A7-ADS-GIS showed
maximum activity at 8.0 and kept about 80% of the maximum
activity even at pH 6.0.   Such high activity at lower pH
region seems to be advantageous for prevention of undesirable

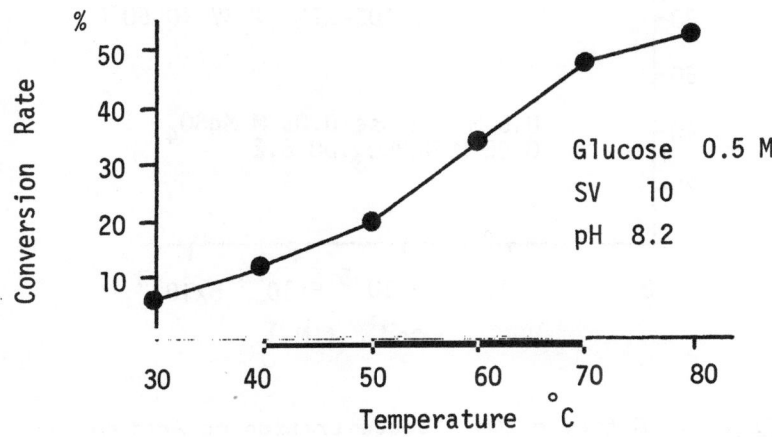

<u>Fig. 2</u>:   Effect of temperature on activity of A7-ADS-GIS

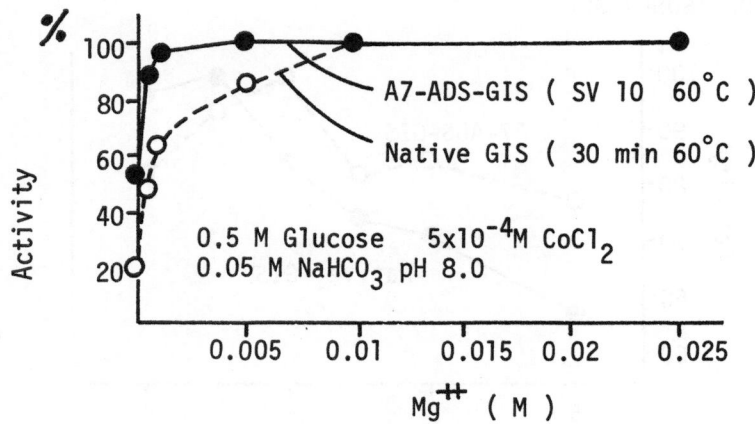

Fig. 3: Effect of Mg^{++} Concentration on Activity

browning reaction during operation. Effect of temperature
on the Duolite A7-ADS-GIS was examined in the temperature
range from 30 to 80°C. Enzyme activity increased in accordance with the increase of temperature in this range as
shown in Figure 2.

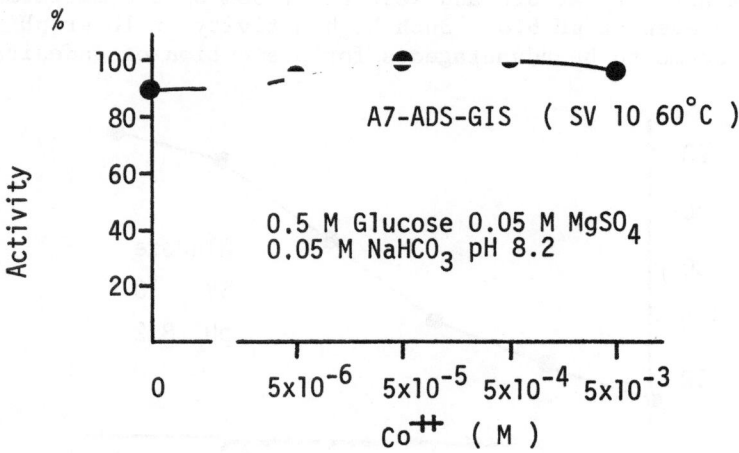

Fig. 4: Effect of Co^{++} Concentration on Activity

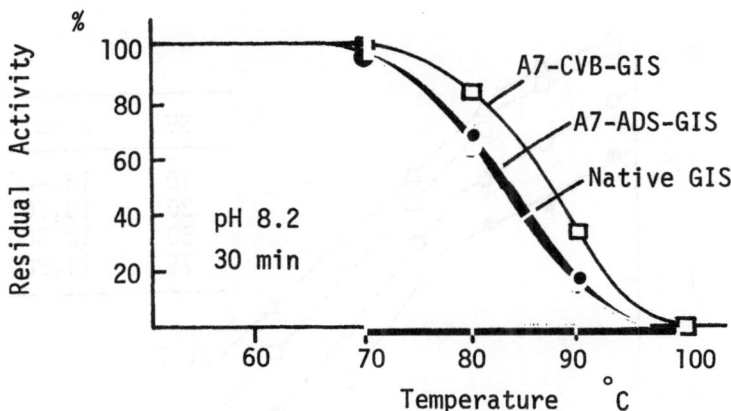

Fig. 5: Heat stability of immobilized GIS

Effect of metal ions. It was noted that native GIS was activated by Mg^{++} and Co^{++} ions (3,9). As shown in Figure 3, in the case of Duolite A7-ADS-GIS, maximum activity was attained by less concentration of Mg^{++} ions than the native enzyme. In Figure 4, it is shown that effect of Co^{++} ions was not significant in the case of Duolite A7-ADS-GIS.

Heat stability of the immobilized enzymes. As shown in Figure 5, all of the native GIS, Duolite A7-ADS-GIS, and Duolite A7-CVB-GIS were quite stable in the heat treatment

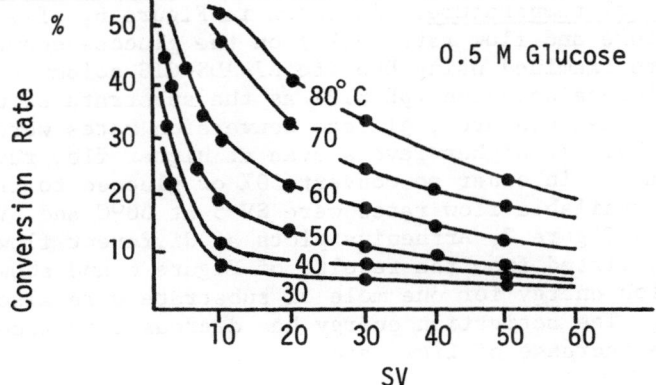

Fig. 6: Effect of temperature and SV on conversion rate

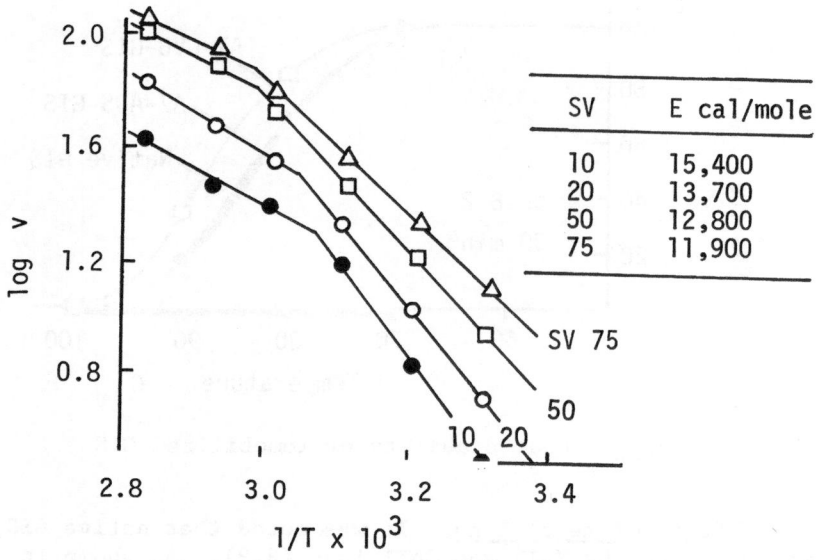

Fig. 7: Arrhenius plot

for 30 minutes below 70°C at pH 8.2. However, at more than 70°C, Duolite A7-CVB-GIS showed better stability than the native GIS and Duolite A7-ADS-GIS.

   Relationship between enzyme activity and physico-chemical factors such as substrate concentration, flow rate and operational temperature. As shown in Figure 6, effects of temperature and flow rate (S.V.) on the glucose conversion rate were examined using Duolite A7-ADS-GIS column and 0.5 M glucose solution (pH 8.2) as the substrate solution. At higher temperatures, glucose conversion rates were usually kept at higher levels even at higher flow rates. For example, in order to convert 50% of glucose to fructose, maximum available flow rates were SV 5 at 60°C and SV at 80°C. In Figure 7, Arrhenius plots at different flow rates were calculated from the results of Figure 6 and shown. Activation energy for one mole of substrate were also calculated. The activation energy has decreased in accordance with the increase of flow rate.

   Effects of glucose concentration and flow rate on the

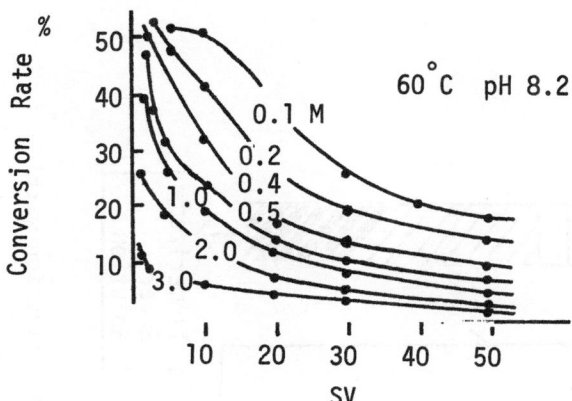

<u>Fig. 8</u>: Effect of glucose concentration and SV on conver-
          sion rate

enzyme activity of Duolite A7-ADS-GIS were examined at 60°C
and pH 8.2. Glucose concentration varied from 0.1 to 3.0
M. And the results are shown in Figure 8. Effects of flow
rate (SV) on the Michaelis constant of Duolite A7-ADS-GIS
were also calculated and shown in Figure 9. The apparent

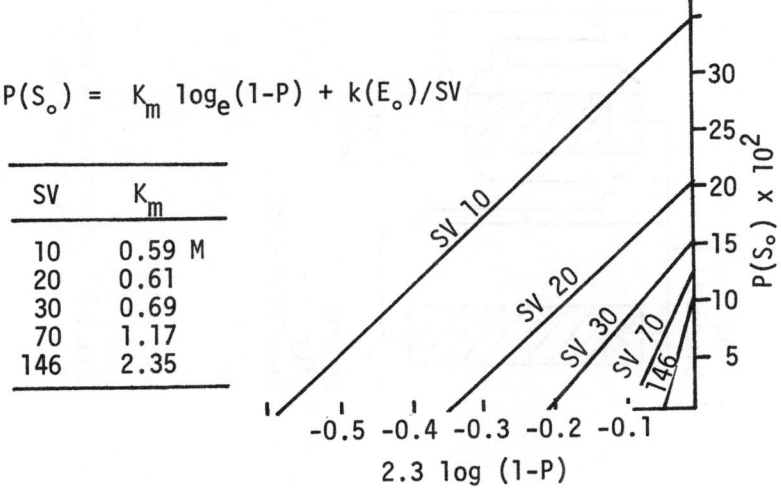

<u>Fig. 9</u>: Effect of SV on the Michaelis constant of A7-ADS-GIS

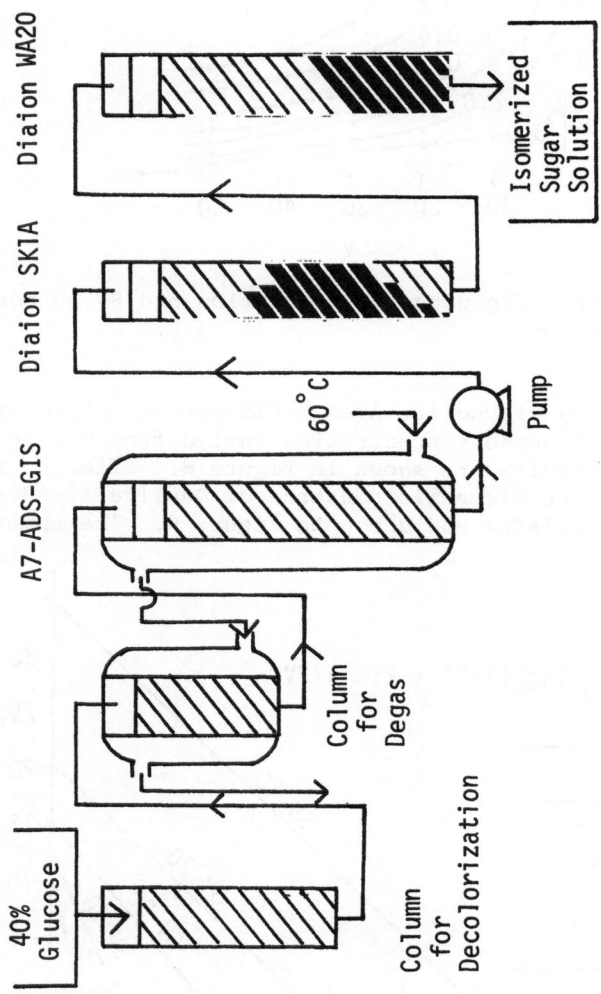

Fig. 10: Continuous operation system for glucose isomerization with Duolite A7-ADS-GIS

Km values increased in accordance with the increase of
flow rate (SV).

Practical glucose concentrations for obtaining 50%
conversion of glucose to fructose were also examined under
a fixed condition (pH 8.2, 60°C, SV 1). Such glucose concentrations were 1 M for Duolite A7-ADS-GIS, and 2.22 M for
Duolite A7-CVB-GIS.

Long-run Continuous Operation of Immobilized GIS

Equipments and operational conditions. Continuous
operation system for glucose isomerization with Duolite
A7-ADS-GIS is shown in Figure 10. Two hundred ml of the
immobilized GIS were packed in the glass column (2.5 x 42
cm D/L) with a jacket in which warm water (60°C) was cir-
culated. Before feeding the substrate solution, the substrate
solution was passed through pretreatment columns which contained each 50 ml of Duolite A7 (bufferized) for eliminating
some impurities and air-bubbles. Forty percent (w/v) glucose solution kept at 60°C was continuously fed through the
pre-treatment and immobilized enzyme columns in series.
The effluent solution from the immobilized enzyme column
was then passed through a cation exchanger (Diaion SK-1A,
H-type) and an anion exchanger (Diaion WA 20, OH-type)
columns successively in order to eliminate salts from the
reaction mixture.

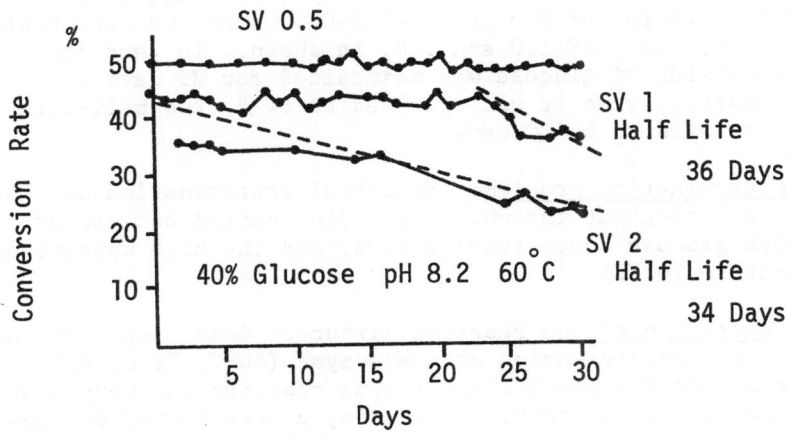

Fig. 11: Continuous operation of A7-ADS-GIS

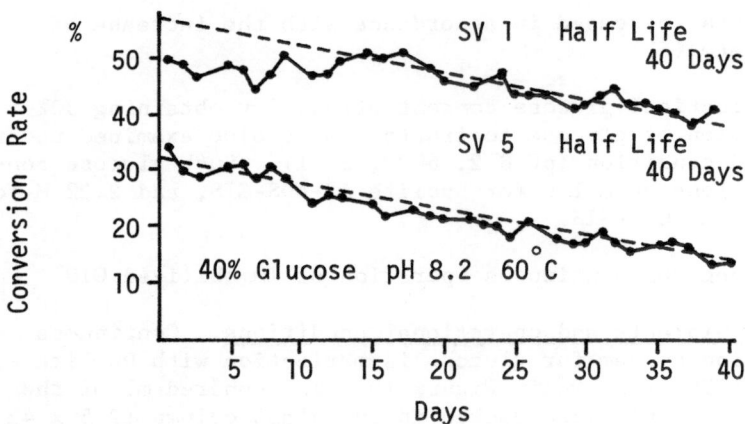

<u>**Fig. 12**</u>:   Continuous operation of A7-CVB-GIS

<u>Half-lives of immobilized enzymes</u>.   In Figure 11, con-
tinuous operation of Duolite A7-ADS-GIS using three different flow rates, i.e., SV 0.5, 1.0 and 2.0, are shown.   At
flow rate, SV 0.5, no significant loss of activity was observed even after 30 days operation, and 50% conversion of
glucose to fructose was maintained during the whole course.
However, when it was operated at a flow rate of SV 2.0,
gradual decrease of the activity was observed and the half-
life of it was calculated to be 34 days.   In Figure 12, continuous operation of Duolite A7-CVB-GIS using two different
flow rates, i.e., SV 1.0 and 5.0, is shown.   In this case,
50% conversion of glucose was maintained for 25 days at SV
1.0.   However, when it was operated at SV 5, the half-life
was calculated to be 40 days.

<u>Contamination problem</u>.   Microbial contamination was not
observed throughout the whole operation period because of
the high glucose concentration (40%) and the high operational
temperature (60°C).

<u>Coloration of the reaction mixture</u>.   Comparing with the
batch type reaction using native enzyme (60°C, 72 hours),
time required for immobilized enzyme reaction was very short,
i.e., for about one hour.   Therefore, coloration of the reaction mixture was very small, i.e., O.D. of the effluent at
410 m μ was about 0.1.

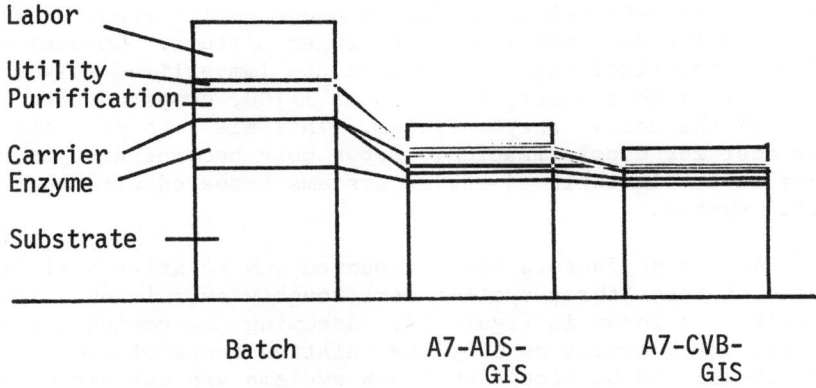

Fig. 13: Comparison of variable cost for isomerization of glucose

DISCUSSION

In order to evaluate the economy of immobilized glucose isomerase process hitherto described, tentative cost estimation was carried out in comparison with the native glucose isomerase process. As the common basis, it was presumed that 50 tons of glucose were converted monthly to isomerized sugar mixture containing 45% fructose. In order to isomerize 1,000 kg of glucose, 21 kg of GIS containing fungal mycelia are required in the batch process using native enzyme. In case of the immobilized enzyme systems 9.8 1 of immobilized enzyme can be obtained from the same amount of mycelia.

In case of Duolite A7-ADS-GIS, 9.8 1 of the immobilized enzyme can isomerize 2,822 kg of glucose within 30 days when it is operated at a flow rate of SV 1 and 60°C using 40% glucose solution as substrate. In case of Duolite A7-CVB-GIS, 9.8 1 of the immobilized enzyme can isomerize 5,644 kg of glucose in a month if it is operated at a flow rate of SV 2 and 60°C. Also, it is presumed that the immobilized enzyme system is operated using 2 columns in series, and an older column is replaced with a fresh column each 30 days in case of Duolite A7-ADS-GIS and each 40 days in case of Duolite A7-CVB-GIS. The requirements of $MgSO_4$ and $CoCl_2$ in the immobilized glucose isomerase systems are 1/10 and 1/20 respectively compared with the native enzyme system. This can decrease the volume of ion exchange resins required for deionization from the isomerized sugar mixture. Coloration of the isomerized sugar mixture in the immobilized enzyme system is much smaller, i.e., about 1/100, compared with that of the native enzyme system. This also can decrease the cost for decolorization. Labor cost becomes also much less in the immobilized enzyme systems compared with the batch system.

All those factors were accounted and relative variable costs of these three systems were roughly calculated. The results are shown in Figure 13. Assuming the variable cost in the batch system as 100, the relative costs of Duolite A7-ADS-GIS and Duolite A7-CVB-GIS systems are calculated as 61.5 and 54 respectively. Of course, indirect cost must be counted. However, such indirect cost may favor the immobilized enzyme systems further.

This cost estimation is only tentative, and must be amended by the fructuation of the prices of every factor. However, it is rather clear that the immobilized enzyme system can decrease significantly the costs for labor, enzyme and purification.

REFERENCES

1. Samejima, H. & Kimura, K., in "Enzyme Engineering", Vol. 2, p. 131, (Eds E.K. Pye and L.B. Wingard, Jr.), Plenum Press, New York & London (1974).

2. Tsumura, S. & Ishikawa, M., J. Food Science and Technology 14: 539 (1967).

3. Strandberg, G.W. & Smiley, K.L, Appl. Microbiol. 21: 588 (1971).

4. Kasumi, T., Kawashima, K. & Tsumura, N., F. Ferment. Technol. 52: 321 (1974).

5. Giovenco, S., Morisi, F. & Pansolli, P., FEBS Letters 36: 57 (1973).

6. Strandberg, G.W. & Smiley, K.L., Biotechnol. Bioeng. 14: 509 (1972).

7. Takasaki, Y. & Kanbayashi, A., Rep. Ferment. Res. Inst. (Japan) 37: 31 (1969).

8. Kay, G. & Crook, E.M., Nature 216: 513 (1967).

9. Tsumura, N. & Sato, T., Agr. Biol. Chem. (Tokyo) 29: 1129 (1965).

IMMOBILIZATION OF ENZYMES BY RADIATION COPOLYMERIZATION OF SYNTHETIC MONOMERS

K. Kawashima and K. Umeda

National Food Research Institute
1-4-12, Shiohama, Koto-ku
Tokyo, Japan

A new method to entrap enzymes in polymer matrix was presented. In the method, ionizing radiation was applied to polymerize synthetic monomer in the presence of enzyme.

A solution containing enzyme and several synthetic monomers was frozen in dry ice-acetone. The frozen mixture was irradiated with Co^{60} at a dose late of 50 - 60 Krad for 5 - 10 min. After the mixture was allowed to stand for a while at room temperature, a spongy immobilized enzyme was obtained. Compared to the former method (1), preparations with much higher activity were obtained. For example, retained activity of α-amylase was 61.6%, alkaline protease 10.6%, neutral protease 16.9%, acid protease 37.9%, glucoamylase 22.1% and catalase 94.1% respectively.

There are many water soluble monomers and polymers which can be polymerized by ionizing radiation in frozen state. By combining these monomers and polymers, various types of polymers can be prepared.

For the immobilization of glycoamylase, the combination of AA (30 g acrylamide, 1.6 g Bis acrylamide, 100 ml water), sodium acrylate (30% w/w), calcium acrylate (30% w/w) and enzyme solution (1:1:2:1 by volume) was found to be the most favorable.

Immobilized glucoamylase showed a shift in pH optimum toward acidic site by about one pH unit. The optimum reaction temperature was slightly higher than that of native glucoamylase. Retained activities were 20 to 46% depending

69

on the enzyme concentration.

The immobilized enzyme prepared by this method had a spongy structure. The preparation has large surface area and the higher retained activities. Compared to the acrylamide polymer (Fig. 1), copolymer had more regular spongy form similar to honeycomb (Fig. 2).

Immobilized enzyme preparation in various shapes such as membranes, bags, tubes and beads (Fig. 3) can be prepared.

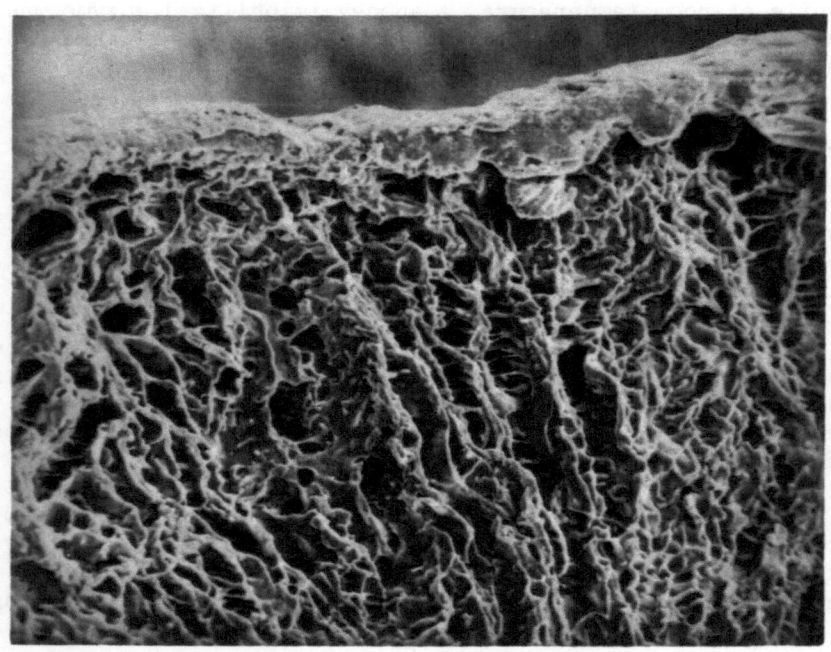

Fig. 1:   Polymer prepared by radiation polymerization of acrylamide. x 270

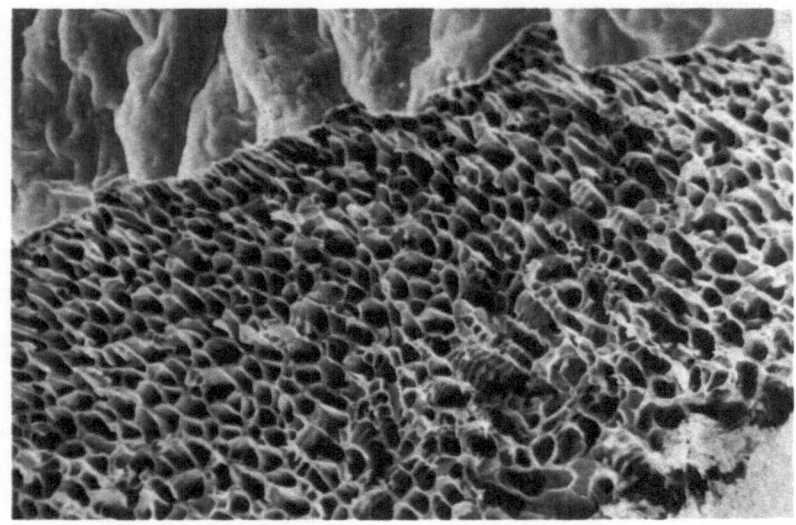

Fig. 2: Polymer prepared by radiation copolymerization of
synthetic monomers. x 270

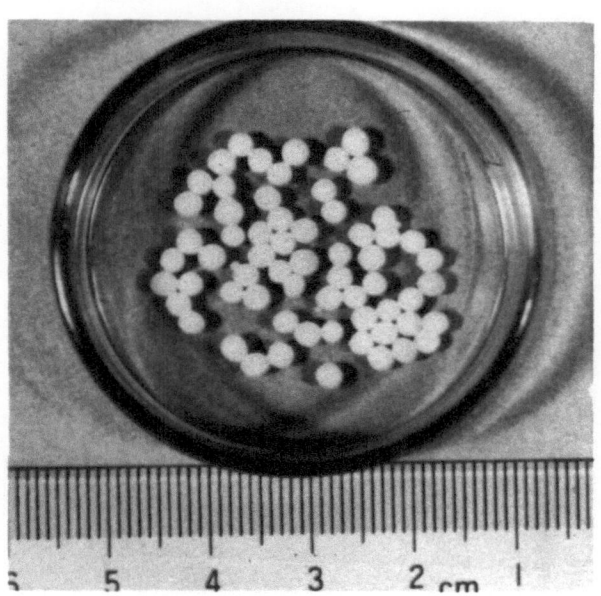

Fig. 3: Bead
shaped immo-
bilized en-
zyme prepared
by radiation
copolymeriza-
tion of syn-
thetic mono-
mers.

REFERENCE

1. Kawashima, K. & Umeda, K., Biotechnol. Bioeng. 16: 609
    (1974).

ACTIVITY OF ENZYME IMMOBILIZED BY MICROENCAPSULATION

Yoshiharu Miura, Kazuhisa Miyamoto, Tomoko Fujii,
Norimasa Takamatsu and Mitsuo Okazaki
Department of Chemical Engineering, Faculty of
Pharmaceutical Sciences, Osaka University,
Suita-shi, Osaka-fu, Japan

INTRODUCTION

It is well known that the immobilization may alter
both the chemical and physical properties of an enzyme,
such as its pH-activity behavior, apparent saturation constant, K_m, substrate specificity and stability toward conformational inactivation, in addition to simply restricting
its gross physical movement. The enzyme activity is generally decreased through immobilization. It is necessary for
obtaining an immobilized enzyme preparation of high activity
to discuss the processes of enzyme immobilization and investigated the factors affecting the activity and yield of
immobilized enzyme. In the present work, we discussed the
processes of enzyme immobilization by microencapsulation
and clarified the factors affecting the activity and yield
of enzyme immobilized.

Among many methods available for enzyme immobilization
(1), microencapsulation has been preferred for medical applications because microencapsulated enzymes or detoxicants
can act without immunological reactions (2,3,4). Studies
on semipermeable microcapsules in this field were pioneered
by T.M.S. Chang who used microcapsulated enzymes or detoxicants of cellular dimensions as "artificial cells" (5).
Preparation of microcapsules containing enzymes and/or other
materials were developed by means of organic phase separa-
tion and interfacial polymerization techniques (6,7,8,9).
Capsules prepared through a secondary emulsion (water-in-
oil-in-water, $W_1/O/W_2$, emulsion) have also been studied by
Kitajima et al (10,11). This method gives microcapsules

without chemical reaction to form the membrane, and can be
expected to produce microencapsulated enzyme in high yield.

The present paper is concerned with microencapsulation
through secondary emulsion. The effect of solute addition
to the two aqueous phase of the emulsion on the capsulation
yield and also on the activity of microencapsulated glucose
oxidase was studied.

EXPERIMENTAL

Semipermeable microcapsules were prepared according to
Kitajima et al (10) with some modifications. Half a ml of
enzyme aqueous solution (5 mg/ml), W_1 phase, was emulsified
in 3 ml of ethyl cellulose and/or polystyrene in benzene
solution, 0 phase. A tissue homogenizer with a teflon pes-
tle was employed for the first emulsification. The emulsification was completed to sufficiently short time (5 or
6 seconds) that the enzyme would not be inactivated by the
heat generated. This emulsion was poured into an agitated
second aqueous phase (40 ml), W_2 phase, containing a suitable emulsifying agent, polyvinyl alcohol (PVA) or sodium
laurylsulfate (SLS). The resulting $W_1/0/W_2$ multiple emulsion was stirred continuously with a magnetic bar in a 100
ml baffled Erlenmeyer flask for the time required to complete the extraction and evaporation of the organic solvent
(usually 2 to 3 hours). The membrane enveloping the W_1
phase was formed during this last operation. Microcapsules
thus obtained were filtered and washed repeatedly with 0.1
M phosphate buffer.

The activities of native and immobilized glucose oxidase were measured by the method presented in the previous
paper (12). Total activity of entrapped glucose oxidase
was measured in the same way after grinding the microcapsules. The concentration of Blue Dextran was determined
from optical density measurement at 630 nm. Radio activity
of $^{32}P-H_3PO_4$ was measured with a liquid scintillation counter (Beckman LS-150, Beckman Instruments, Inc., U.S.A.)

YIELD OF MICROENCAPSULATION

For successful microencapsulation of the enzyme by the
secondary emulsion method, the following conditions should
be satisfied: 1.) the first and second emulsion should be

stable during the microencapsulation, 2.) evaporation of
organic solvent must proceed at an appropriate rate, 3.)
organic solvent, which is in contact with enzyme solution,
must not inactivate the enzyme. The second condition was
necessary because the formation of a membrane was observed
at the free surface when rapid evaporation was performed.
Benzene was found to be a good solvent which satisfied the
last two conditions. Surface aeration was an effective
control of the evaporation rate. Ethyl cellulose in benzene solution gave a very stable W_1/O emulsion without an
emulsifying agent. The stability of the secondary emulsion
seemed to be the most important factor affecting capsulation. Enhanced stability was obtained by the addition of
polyvinyl alcohol (PVA) or sodium laurylsulfate (SLS) to
the external aqueous phase. Many of these observations
agreed with those reported by Kitajima et al (10) in their
studies on microencapsulation of catalase, urease and other
enzyme preparations.

The relation between PVA concentration in the external
aqueous phase and the capsulation yield was obtained, using
0.1 M phosphate buffer solution and deionized water as the
external phase. The results are shown in Fig. 1, where the
capsulation yield is defined as the ratio of glucose oxidase activity in the microcapsules to that of the total enzyme used. The yield was constant over the PVA concentration range studied. PVA dissolved in 0.1 M phosphate buffer gave a higher capsulation yield, in contrast to PVA
dissolved in deionized water.

The effect of SLS concentration of the capsulation
yield is shown in Fig. 2. Constant yield was obtained when
SLS was dissolved in deionized water, but the presence of
$(NH_4)_2SO_4$ altered the SLS concentration effect: the capsulation yield decreased as SLS concentration in 0.1 M $(NH_4)_2SO_4$
solution increased. The salt may affect the surface activity
of SLS. Interfacial tension was thought to control the capsulation since microencapsulation did not occur in the absence of the emulsifying agent in the external aqueous phase.
To check the possibility of marked change in interfacial
tension on addition of $(NH_4)_2SO_4$, the surface tensions of
the two systems were measured with a du Noüy tension meter.
Results are shown in Fig. 3. According to Antonoff's law,
the difference between the surface tensions of the aqueous
solution and the organic solvent is assumed to be the inter-

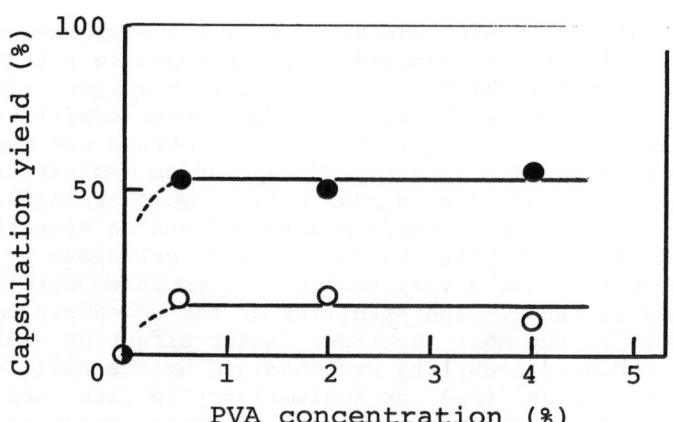

Fig. 1: Effect of polyvinyl alcohol concentration on the capsulation yield.
PVA dissolved in deionized water (O) or in 0.1 M phosphate buffer solution (●) was used as the external aqueous phase. The capsulation yield was based on the entrapped activity of glucose oxidase.

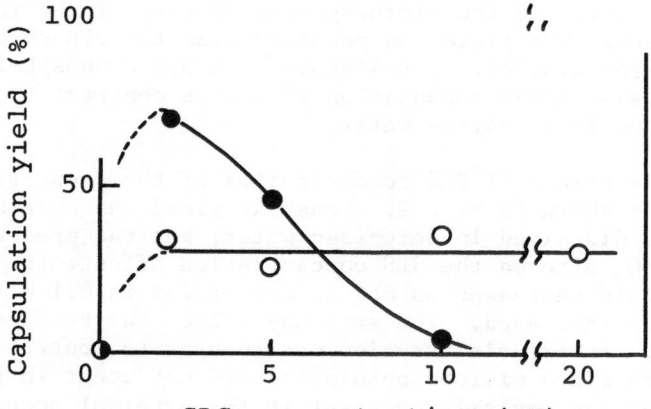

Fig. 2: Effect of sodium laurylsulfate concentration on the capsulation yield.
SLS dissolved in deionized water (O) or in 0.1 M $(NH_4)_2SO_4$ solution (●) was used as the external aqueous phase. The capsulation yield was based on the entrapped activity of glucose oxidase.

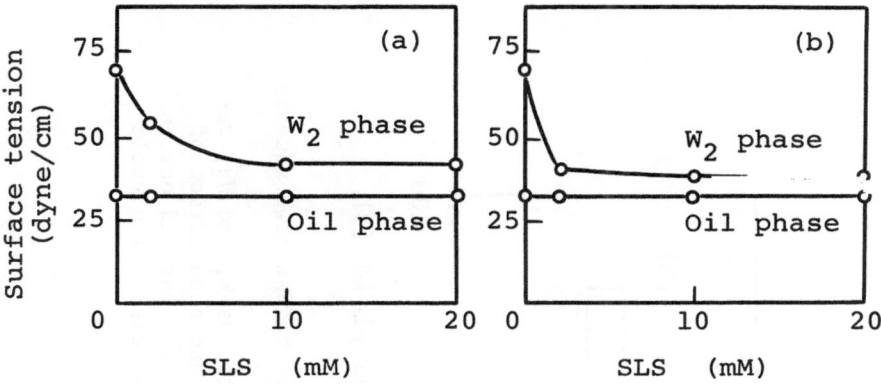

Fig. 3: Surface tensions of benzene and aqueous phases.
Both phases were pre-equilibrated with each other.
SLS dissolved in deionized water (a) or in 0.1 M
$(NH_4)_2SO_4$ solution (b) was used in correspondence
with the experiments shown in Fig. 2.

facial tension. A slight decrease in Antonoff's interfacial tension was observed when $(NH_4)_2SO_4$ was added to the
water phase, but this could not be correlated with the
marked change in the capsulation yield shown in Fig. 2.

As indicated above, the addition of salt to the external aqueous phase gave a higher yield. Next, the effect of
the difference in the salt concentration of the two aqueous
phases was investigated, using Blue Dextran as an indicator
for obtaining the capsulation yield. Fig. 4(a) shows the
capsulation yields of four preparations, in which SLS was
used to stabilize the secondary emulsion. The highest yield
was obtained by the addition of $(NH_4)_2SO_4$ to only the external aqueous phase. Conversely, on addition of $(NH_4)_2SO_4$ to
the internal phase alone, the indicator could scarcely be
encapsulated. The same features were demonstrated in another
set of experiments as shown in Fig. 4(b), in which the secondary emulsifications were carried out in 1% PVA solution.

When sucrose instead of $(NH_4)_2SO_4$ was mixed into the
internal, external, or both aqueous phase, results similar
to those for $(NH_4)_2SO_4$ were obtained, as shown in Fig. 5.
The addition of sucrose to the external SLS solution gave

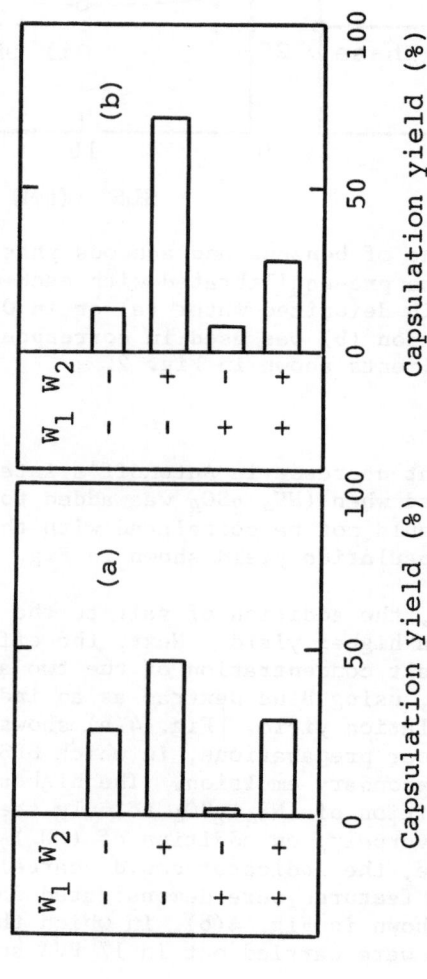

<u>Fig. 4:</u> Effect of the difference in $(NH_4)_2SO_4$ concentration between two aqueous phases on the capsulation yield.
The internal and external aqueous phases are denoted as W_1 and W_2, respectively. Symbols, + and −, represent addition and nonaddition of $(NH_4)_2SO_4$ to that phase. Secondary emulsification was carried out in 5 mM SLS solution (a) or 1% PVA solution (b). The capsulation yield was based on the entrapped amount of Blue Dextran.

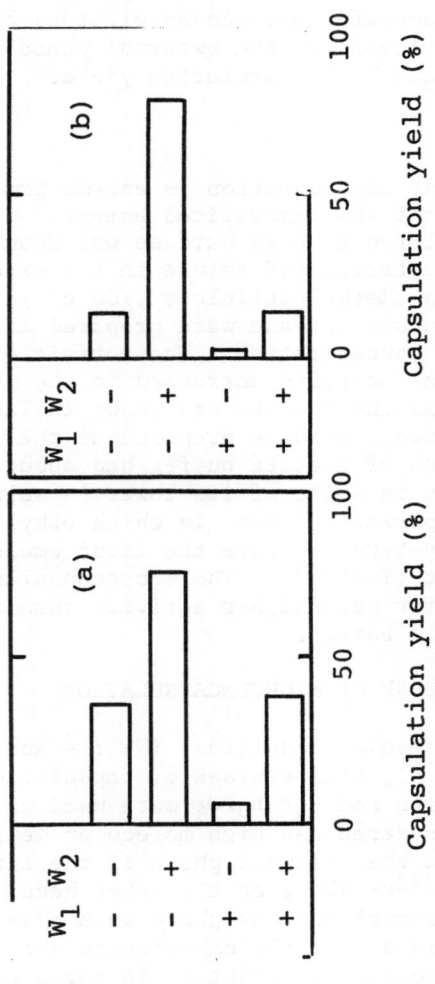

<u>Fig. 5:</u> Effect on the capsulation yield of the difference in sucrose concentration between the two aqueous phases. Secondary emulsification was carried out in 10 mM SLS solution (a) or in 1% PVA solution (b). The capsulation yield was based on the entrapped amount of Blue Dextran. Symbols are as shown in Fig. 4.

high yield of 60 to 70% regardless of SLS concentration,
in contrast to the case of $(NH_4)_2SO_4$ addition as shown in
Fig. 2. The effect of sucrose on the surface activity of
SLS may differ from that of $(NH_4)_2SO_4$. Figure 6 shows that
the addition of a solute to the external phase alone was
not essential for successful microencapsulation, but that
higher concentration levels in the external phase than in
the internal phase gave high capsulation yield.

ACTIVITY

Another important consideration in enzyme immobilization is the activity of the immobilized enzyme. The activity of microencapsulated glucose oxidase was found to be
affected by the concentration of solute in the external
aqueous phase. Using 2%ethyl cellulose (100 cps) in benzene,
microencapsulated glucose oxidase were prepared in acetate
buffers of different concentration. The activities of the
microcapsule and total activity entrapped in the microcapsule were measured and the results are shown in Table 1.
Microencapsulated glucose oxidase prepared at the lower
external concentration of acetate buffer had about four
times higher activity in spite of its lower capsulation
yield. A similar experimental run, in which ethyl cellulose
was blended with polystyrene before the first emulsification, was carried out (Table 2). The microcapsules prepared
in 0.1 M acetate buffer gave higher activity than those prepared in 0.5 M acetate buffer.

THE COURSE OF MICROENCAPSULATION

For two cases of solute addition, $(W_1 : -$ and $W_2 : +)$
and $(W_1 : +$ and $W_2 : -)$, time courses of capsulation were
followed. Blue Dextran and $^{32}P-H_3PO_4$ were used as indicators. Because Blue Dextran has high molecular weight, it
cannot be detected in the external phase if the first emulsion is not broken. $^{32}P-H_3PO_4$, on the other hand, can diffuse out from the internal aqueous phase when a semipermeable membrane is formed during the capsulation process.
$^{32}P-H_3PO_4$ was used because its addition in small amounts to
the internal aqueous phase did not disturb the capsulation,
and its appearance in the external solution could be detected with high accuracy.

Table 1. Effect of the solute concentration
in the external aqueous phase on the activity
of ethyl cellulose microencapsulated glucose
oxidase.

Concn. of acetate buffer in W_2 phase (M)	Activity of microcapsule (%)*	Capsulation yield (%)
0.1	4.3	37
0.5	0.97	43

* The activity of microcapsule was expressed as
per cent of the activity of the total enzyme used.

Table 2. Effect of the solute concentration on the
activity of microencapsulated glucose oxidase
using blended polymer.

Ratio of ethyl cellulose to polystyrene**	Concn. of acetate buffer in W_2 phase (M)	Activity of microcapsule (%)*	Capsulation yield (%)
1 : 1	0.1	5.3	42
	0.5	3.6	35
1 : 5	0.1	10.	31
	0.5	2.5	32

*   The activity of microcapsule was expressed as per cent
of the activity of the total enzyme used.

** Total polymer concentration was kept constant (4 %).

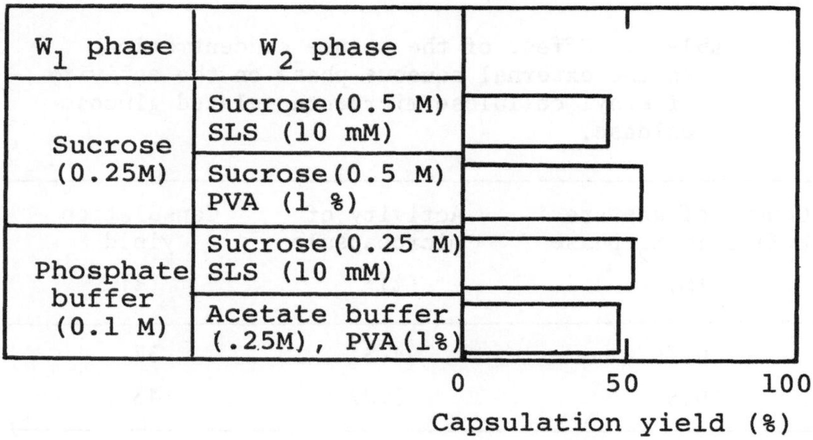

W_1 phase	W_2 phase	
Sucrose (0.25M)	Sucrose (0.5 M) SLS (10 mM)	
	Sucrose (0.5 M) PVA (1 %)	
Phosphate buffer (0.1 M)	Sucrose (0.25 M) SLS (10 mM)	
	Acetate buffer (.25M), PVA(1%)	

```
                              0        50        100
                              Capsulation yield (%)
```

Fig. 6: Effect on the capsulation yield of the difference in solute concentration between the two aqueous phases. For each preparation, the internal and external aqueous phases had the composition shown on the left. The capsulation yield was based on the entrapped amount of Blue Dextran.

Figure 7 represents the time courses of the two pre-parations; one, shown in Fig. 7(a), was emulsified in the second aqueous phase of 0.5 M acetate buffer solution; the other, shown in Fig. 7(b), was emulsified so as to contain 0.1 M acetate buffer together with the two indicators in the internal aqueous phase. In the former case, the leakage of Blue Dextran was small; after 20 hours only 30% of the amount added was detected in the external phase. The leakage of the smaller indicator, ^{32}P-H_3PO_4, was greater than that of the larger one, Blue Dextran, even in the early stages of the capsulation process. This observation leads to the conclusion that oil droplets incorporating finer aqueous droplets become semipermeable early in the process. In the second case, the leakages of Blue Dextran and ^{32}P-H_3PO_4 rapidly increased after lag periods of 30 to 40 minutes. Both indicators appeared completely in the external phase after 20 hours.

Polystyrene instead of ethyl cellulose dissolved in benzene did not give a stable W_1/O emulsion. When this emulsion was poured into the second aqueous phase, leakage of

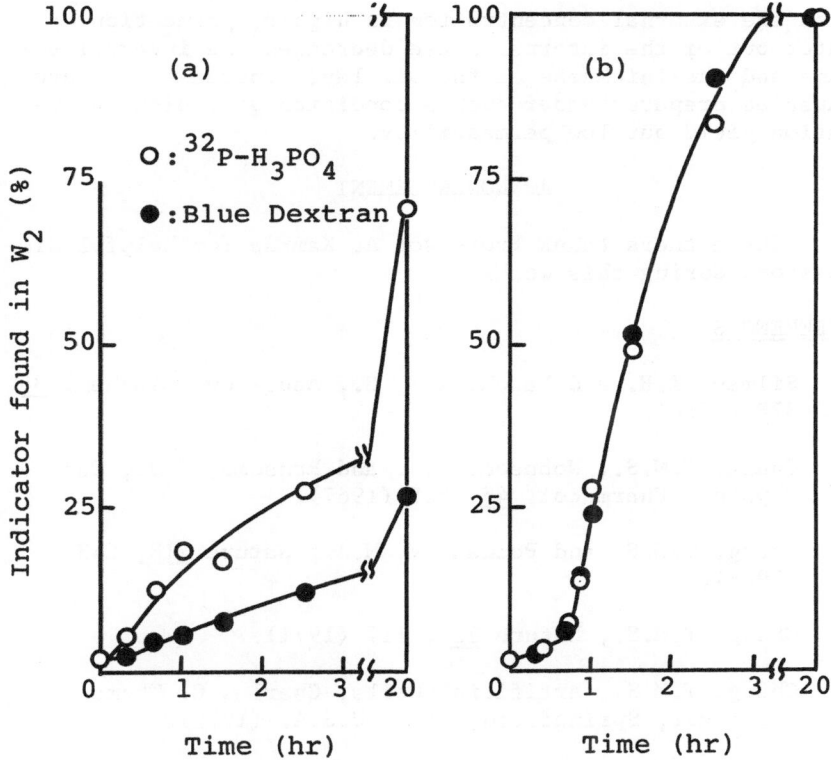

Fig. 7: Time courses of the microencapsulation. Two percent PVA in 0.5 M acetate buffer solution (a) or
in deionized water (b) was used as the external
aqueous phase. Blue Dextran and $^{32}P-H_3PO_4$ were
added to the internal aqueous phase and their appearance phase was followed.

the indicators was detected immediately.

CONCLUSIONS

Oil droplets become semipermeable prior to completion
of microencapsulation. If the concentration of solutes in
the internal aqueous phase is higher than that in the external, permeation of water into the internal phase brings
about an increase in the internal volume, the oil layer becomes thinner and thinner, and is finally broken. Conversely,

when the external concentration is higher, permeation of water out of the internal phase decreases the internal volume and the thickness of the oil layer increases. Microcapsules prepared under such a condition give high capsulation yield but low permeability.

ACKNOWLEDGEMENT

The authors thank Professor A. Kamada for helpful discussions during this work.

REFERENCES

1. Silman, I.H. and Katchalski, E., Ann. Rev. Biochem. 35, 873 (1966).

2. Chang, T.M.S., Johnson, L.J. and Ransome, O.J., Can. J. Physiol. Pharmacol. 45, 705 (1967).

3. Chang, T.M.S. and Poznansky, M.J., Nature 218, 243 (1968).

4. Chang, T.M.S., Nature 229, 117 (1971).

5. Chang, T.M.S., Artificial Cells, Charles C. Thomas Publisher, Springfield, Ill., U.S.A. (1972).

6. Chang, T.M.S., Science 146, 524 (1964).

7. Chang, T.M.S., MacIntosh, F.C. and Mason, S.G., Can. J. Physiol. Pharmacol. 44, 115 (1966).

8. Suzuki, S., Kondo, T. and Mason, S.G., Chem. Pharm. Bull (Tokyo), 16, 1629 (1968).

9. Mori, T., Tosa, T. and Chibata, I., Biochim. Biophys. Acta 321, 653 (1973).

10. Kitajima, M., Miyamoto, S. and Kondo, A., Kogyo Kagaku Zasshi, 72, 493 (1969).

11. Kitajima, M. and Kondo, A., Bull. Chem. Soc. Japan 44, 3201 (1971).

12. Miyamoto, K., Fujii, T., and Miura, Y., J. Ferment. Technol. 49, 556 (1971).

ARTIFICIAL ENZYME MEMBRANES AND ENZYME REACTORS

Daniel Thomas and Gerard Gellf

Laboratoire de Technologie Enzymatique ERA n°
338 du C.N.R.S.U.T.C. B.P. 233 60206 Compiegne

The aim of this paper is to describe shortly the work performed by the Laboratory of Enzyme Technology in the University of Technology of Compiegne in the field of immobilized enzymes. The results are classified in three parts:

-Kinetic and Transport phenomena in artificial enzyme membranes.
-Packed and Fluidized bed enzyme reactors.
-Analytical and Medical applications.

Numerous papers were devoted to immobilized enzymes and are listed in the "Immobilized Enzyme Compendium" edited by Corning Glass" (1).

I. KINETIC AND TRANSPORT PHENOMENA IN ARTIFICIAL ENZYME MEMBRANES

Within the living cell, the great majority of the enzymes are attached to membrane structures or contained in cell organelles. When enzymes are isolated, they are removed from their natural state and quite often are highly unstable. The artificial binding of enzymes into membranes makes possible study of the interaction between diffusion and enzyme reaction within a well-defined context.

In this way Goldman et al (2,3,4,5) incorporated papain and phosphatase in collodion membranes and other authors (6,7) have done likewise with several enzymes in cellophane membranes. It is important to note the work of Suzuki (8).

These methods produce active artificial membranes but the active site distribution is not well-defined and the theoretical treatment is difficult.

In order to give an homogenous distribution of enzyme molecules inside the membrane, it was necessary to synthesize the membrane and to incorporate the enzymes at the same time. The co-crosslinking of enzyme molecules with an inert protein appears to be a proper solution. Purely active proteic films were created by using this procedure (11,12). These artificial enzyme membranes can be used in the study of heterogeneous enzyme kinetics and for modeling biological membranes. The phenomena in the enzyme membranes can be classified in two parts.

1.) The Effect of the Composition and Structure of the Membrane itself on the Enzyme Membrane Behavior.

In reference to this, it is possible to describe examples of facilitated transport and active transport.

Facilitated transport of CO_2 with a membrane bearing carbonic anhydrase was described by Broun et al (13). In this system a hydrophobic membrane separates two compartments containing buffer solutions. This membrane is permeable to gases like CO_2, but impermeable to water and electrolytes. CO_2 diffusion velocity was measured for a membrane without enzyme; with grafted enzyme on one side and with grafted enzyme on both sides. With enzyme the apparent permeability increased two times and four times respectively The concentration profiles of CO_2 with and without enzyme can explain this phenomenon.

An active glucose transport effect occurs with a bienzyme membrane composed of two active protein layers and two selective films (14,15). The active enzyme films carry, respectively, hexokinase and phosphatase co-crosslinked with an inert protein. Both are impregnated with A.T.P. and covered on their external sides by two selective films permeable to glucose, but impermeable to glucose-6-phospate. In this asymmetrical membrane, glucose is temporarily phosphorylated and the system behaves chemically as a simple A.T.Pase. In the first layer glucose is a substrate, and G-6 P diffuses along its own concentration gradient into the second layer where glucose is a product. The glucose

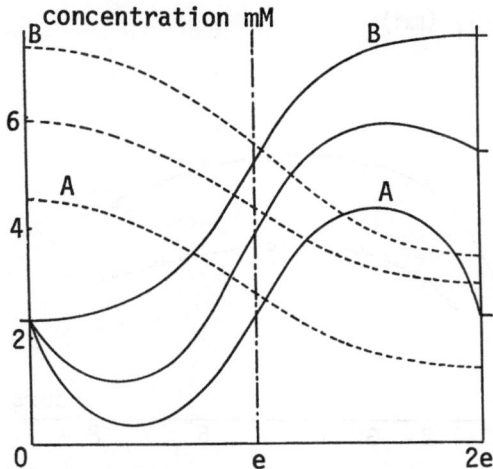

Fig. 1: Substrate and product concentration profiles in a
         bienzyme membrane with a hexokinase-phosphatase
         double-layer system. Glucose concentration pro-
         files are given in _____; glucose-6 phosphate
         concentration profiles in .... When S_1=
         S_2; the glucose profile is given by the curve A.
         When S_1 is smaller than S_2 and $J_1=J_2=0$, the glu-
         cose profile is shown by B. The third glucose
         profile shows an active transport effect.

and glucose-6 phosphate concentration profiles explain how
this diffusion reaction is converted into a pumping phenomenon (Fig. 1).

    When the concentration in the donor compartment remains
constant during the experiment (Fig. 2), the concentration
in the receptor compartment increases regularly at first then
attains a plateau. For an ideal system with the thickness
of a biological membrane a maximum increase in concentration of 130 times can be predicted by computer simulation.
This system gives an experimental physico-chemical example
of a transformation of scaler chemical energy into a "vec-
torial catalysis effect".

2.) The Effect of the Reactant Concentration Distributions
    on the Enzyme Membrane Behavior.

    First of all, the behavior of the enzymes in the mem-

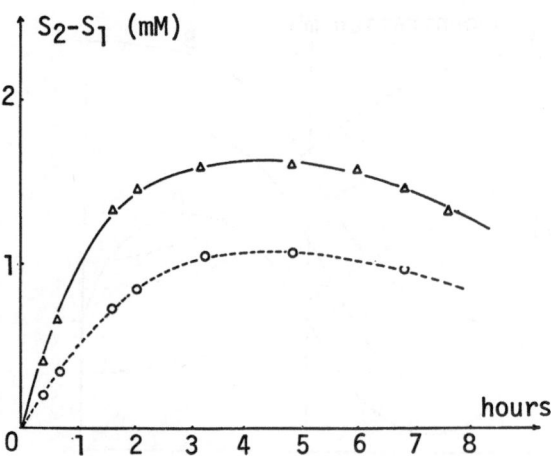

<u>Fig. 2</u>:   Glucose concentration difference between the two
            compartments as a function of time (hours) in the
            described system. Each curve corresponds to one
            value of initial glucose concentration.

brane differs markedly from the behavior of the unbound
enzymes in solution. It is pertinent to note that the
milieu in which the enzyme bound to a membrane acts might
be determined not only by the composition and structure of
the membrane itself but also by the local concentration
distribution of substrate and products. The micro environment in the membranes is the result of a balance between the
flow of matter and enzyme reactions. The substrate and product concentrations in the membrane differ from point to
point across the membrane and also from those at the outer
solution. By electron microscopy this was experimentally
demonstrated beyond doubt with the DAB-peroxidase system by
Barbotin and Thomas (16). The effects of these profiles
were studied with monoenzyme membranes by Goldman et al
(2-5) and Thomas et al (11). A study of activity as a function of inhibitor concentration for enzyme membrane with
different Thiele moduli was done. Enzyme in a membrane is
less sensitive to the effect of inhibitor. In contrast, the
effect of the reaction reversibility on an enzyme reaction
is increased by the diffusion-limitation. These problems
were studied experimentally and by simulation on computer
(17). The profiles are still more important when non linearity of the enzyme reactions can produce memory, oscil-

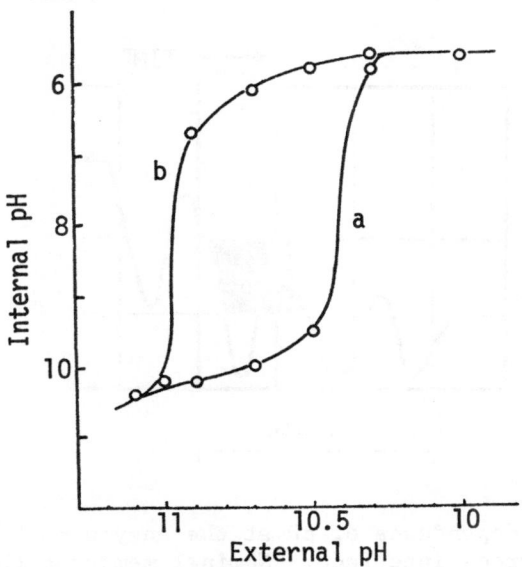

<u>Fig. 3</u>: Papain membrane-coated electrode. Stationary
          state description of internal pH as a function of
          increasing (1) and decreasing (2) the external pH
          values. The experiment started with decreasing
          the pH and then increasing. (See text for details).

lation and spontaneous structuration.

Memory in enzyme membranes (18)

A coating, bearing one enzyme (papain) is produced on
the surface of a glass pH electrode by the method previously
introduced (co-crosslinking). The papain reaction decreases
the pH, and the pH-activity variation gives an autocatalytic
effect for pH values greater than the optimum; under zero
order kinetics for the substrate (Benzoyl Arginine Ethyl
Ester) the pH inside the membrane is studied as a function
of the pH in the bulk solution in which the electrode is
immersed (Fig. 3). An hysteresis effect is observed and
<u>the enzyme reaction rate depends not only on the metabolite
concentrations but also on the history of the system</u>.

Oscillatory behavior of enzyme membranes (19)

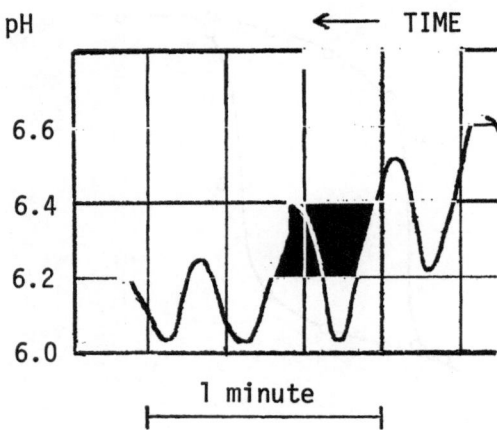

<u>Fig. 4</u>:   Time-dependence of pH at the enzyme membrane-glass
             electrode interface. Nominal membrane thickness
             10 , gentle stirring. Substrate concentration and
             pH in the bulk solution were initially 4.5 mM and
             9.3 respectively.

When the same kind of electrode is introduced in a
solution with a high pH (i.e. pH=10) and a lower substrate
concentration (under first kinetics), an oscillation in
time of the measured pH inside the membrane, spontaneously
occurs, (Fig. 4). This enzyme which has been extensively
studied, does not give oscillation for any conditions of
pH and substrate concentration. The period of oscillation
is around one half minute, and the oscillation is abolished
by introducing of an enzyme inhibitor. The phenomenon can
be explained by the autocatalytic effect and by a feed-back
action of OH^- diffusing in from the outside solution. The
diffusion of this ion is quicker than the diffusion of the
substrate. There is a qualitative agreement between the
computer simulation and the experimental results.

Spontaneous structuration in enzyme membrane (20)

An artificial membrane bearing two different enzymes
(glucose oxidase and urease) in a spatially homogeneous
fashion, is produced by using the method previously described. The glucose oxidase reaction decreases the pH, and

the urease increases the pH. The pH activity profiles show
an autocatalytic effect for the glucose oxidase in the range
of pH values greater than the optimum, and for the urease,
smaller than the optimum pH. When the two enzymes are mixed
together the global pH-variation is zero for one well-defined
pH value.

The active membrane separates two compartments and it
is possible to get this pH-value throughout the system, in
presence of the two substrates, by the transient use of a
buffer. The pH values outside are controlled and H^+ fluxes
measured by pH-stat systems. After small asymmetrical perturbations of the pH values at the boundaries (0.05), an
inhomogeneous pH distribution arises spontaneously inside
the membrane. The initial perturbations are amplified and
the pH values in the compartments tend to evolve in opposite
directions. The H^+ fluxes entering and leaving the membrane
can be determined by pH-stat measurements. If the boundary
pH values are not maintained constant by a pH-stat, the system evolves to a new stationary state characterized by a pH
gradient of 2 pH units across the membrane.

II. PACKED AND FLUIDIZED BED ENZYME REACTORS

1.) Packed Bed Reactors

Several enzymes (oxido-reductases, transferases, ly-
ases and hydrolases) were immobilized by a co-crosslinking
procedure into purely proteic particles which could be
packed into columns (12). By using this method (cheap and
easy to perform) an important part of the initial enzyme
activity remained available after immobilization: ratios
from 30% to 80% were currently observed. Such particles
were packed into a thermostated column which is continuously
flowed through by the substrate. The outlet product concentration was measured as a function of the following
parameters: flow rate, inlet substrate concentration, enzyme activity and reaction mechanism: michaelian, competitive inhibition, two-substrate system with inhibition by
one of the products. Kinetic behavior of immobilized enzymes when used in columns depends on three phenomena:
enzyme reaction, metabolite diffusion inside particles and
metabolite convection in solution.

As a simplification tool, the following kinetic model-
izations may be distinguished and characterized:

"Plug flow" model (abreviation: P F model)

-one phase
-one axis (flow rate direction)
-substrate mass balance involves convection and enzyme
reaction
a1) Steady state hypothesis

This steady state hypothesis ($\partial S/\partial t = 0$) implies that
the whole column is in steady state. In this particular
case, the equation presents a simple analytical solution
(9,10). Validation of this solution by comparison with
experimental results may be tested by an appropriate linear
transformation. It was observed by several authors (9,10,
21,22) that this validation occurs only when "apparent"
values of the kinetic constants ($V'm$, $K'm$, K'_I) were used
instead of the experimental values, and that these apparent
constants were varying with the flow rate but independent
of the substrate concentration.

a2) Transient conditions

A steady state kinetic study implies the hypothesis of
a constant substrate concentration as a function of time at
any point on the column. This hypothesis of constancy for
the concentration profiles necessitates constancy, during
the whole experiments for:

-kinetic parameters (Vm and Km)
-flow rate of substrate
-ingoing substrate and effectors concentrations (activators or inhibitors).

Insofar as monoenzyme columns are concerned the hypothesis of successive steady state can be used (discrete
variations of parameters) but for a multienzyme sequential
system where intermediary products can have an activating
or an inhibiting effect, it becomes necessary to take into
account the reaction of the system to concentration variations of the components. Kinetic knowledge of transient
behaviors therefore becomes necessary. For the same reasons, regulation and control by injected activators, inhibitors, or consubstrates can only be studied in transient
conditions. In this case, P F equation was solved by num-
eric calculations on a digital computer (22). The apparent

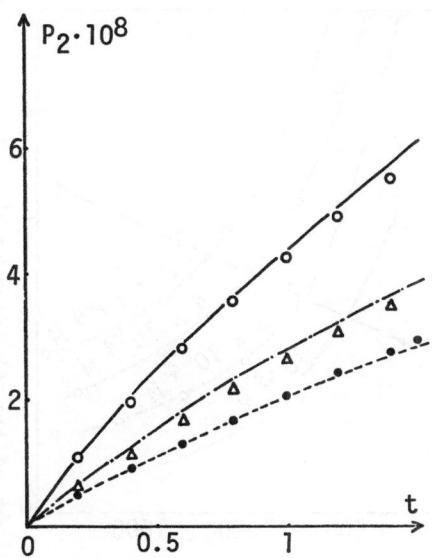

Fig. 5: Output product concentration vs. time. Comparison
between experimental points (O △ ●) and calculated curves (___, _ _, _ _). The values of the
ingoing substrate concentration gradient being:
$S_1 = 0.6 \times 10^{-7} t$ (--- and ●); $S_2 = 0.8 \times 10^{-7} t$ (- -
and ▲); $S_3 = 1.406 \times 10^{-7} t$ (___ and O). (S and P_2
in moles cm^{-3}, t in hours).

constants used (V'm and K'm) were experimentally determined
at stationary state for several values of the flow rate.
The evolution of both substrate and product concentrations
profiles inside the column and effluent concentrations were
calculated. The gap between experimental and predicted
values (Fig. 5) is equal to or less than 5%, value which
does not exceed experimental errors in the determination of
V'm and K'm. The quantitative coincidence between calculated and experimental results for transient conditions (ingoing concentrations varying from 0 to 1. 4 x 10^{-6} moles
cm^{-3}) is another verification of this hypothesis: for a
given flow rate, determined V'm and K'm in steady state
conditions may be considered as characteristics of the kinetic behavior of the column.

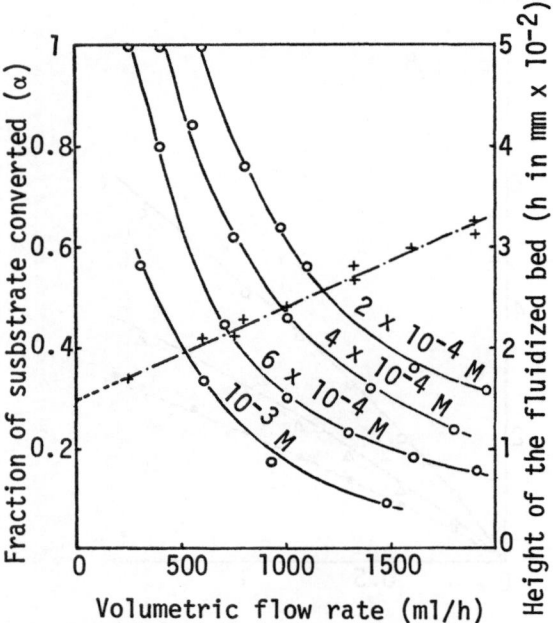

<u>Fig. 6</u>: Fraction of substrate converted (___O___) and
             height of the fluidized bed (___+___) vs.
             columetric flow rate. Molarities of inlet sub-
             strate solutions are indicated on the curves.

        b) Diffusion-Convection-Reaction model (abreviations:
           D.C.R. model)

        -two axis (flow rate direction and the perpendicular
one)
        -two phase: bulk solution (convection and diffusion
of metabolites) and enzyme insoluble phase (diffusion of
metabolites and enzyme reaction).

        In this case, the equations can be solved numerically
only (24), so that transient and steady states were treated
simultaneously. With this modelization, it was possible to
obtain an agreement with the experimental results, by using
only true values for the kinetic parameters (Vm, Km, Ki).
But another point is important to mention: if at these calculated values we applied the linear transformation of P F
model, they gave rise to the same behavior than experimental

values: linearity is observed, but the kinetic constants
calculated from their slopes and initial ordinates (K'm and
V'm) depend on the flow rate (24).

2.) Fluidized Bed Enzyme Reactors (23)

        Invertase, glucose oxidase and papain were immobilized
on magnetic iron-nickel particles (100 μin diameter) by a
co-crosslinking between the enzyme and an inactive protein
(bovine albumin) with glutaraldehyde. Active magnetic particles bearing papain activity were used in a fluidized bed
reactor: 10g of particles were poured in a thermostated
chromatographic column (10mm internal diameter, 50cm length).
In order to retain the finest particles, the top of the column was surrounded by a circular magnet. Substrate solutions at various concentrations were flowed through the bottom of the column by a peristaltic pump. The fraction of
substrate was determined as a function of the flow rates and
of the inlet substrate concentrations. The height of the
fluidized bed was varying linearly with flow rate and the
minimum fluidizing velocity was found to be about 250 ml^{-1}
(Fig. 6). Carry over of fine particles occurred at high
flow rates but was efficiently stopped by the magnet. The
absence of iron and enzyme leakage was cheked in the outlet
solutions. Further work along this line is in progress, including long term assays, kinetic study and modelization.

III.   ANALYTICAL AND MEDICAL APPLICATIONS

1.) Analytical Applications

        One of the most interesting uses of enzyme membranes
has been as the active element of an electrode (25). Lysine,
Tyrosine and Phenylalanine electrodes were prepared by using
Lysine decarboxylase, Tyrosine decarboxylase and Phenylalanine decarboxylase membranes (25). These membranes were rea-
lized by the co-crosslinking method previously described.
The electrode consists of a combination of pH glass and reference electrode and a gas permeable membrane according to
the classical CO_2 electrode. Between the glass electrode
and the gas membrane there is a layer of sodium bicarbonate
solution. The active membrane takes place along the external side of the gas impermeable membrane. When these electrodes are in contact with sample solutions containing their
respective amino acid, the CO_2 is produced inside the membrane. A local concentration appears which is monitoring

by the CO_2 electrode. This local concentration is linked
to the amino acid concentration through the enzyme kinetics. There is a well defined relationship between the
pCO_2 measurement and the amino acid concentration in the
sample solution.

As far as the selectivity is concerned the CO_2 electrode is undoubtedly far superior to any other known electrode because of its gas permeable membrane.

2.) Medical Applications

By grafting enzymes on soluble polymers of albumin
molecules it is possible to product stabilized soluble
enzyme activities allowing intraveneous injection.

These active soluble polymers offer a solution for
the immunological problems of enzymotherapy (26).

REFERENCES

1. Compedium of references: "Immobilized Enzymes", Ed.
    Corning Glass-Corning, N.Y.

2. Goldman, R., Silman, H.I., Caplan, S.R. Kedem, O. &
    Katchalski, E., Science, 150:758 (1965).

3. Goldman, R., Kedem, O., Silman, H.I., Caplan, S.R. &
    Katchalski, E., Biochemistry 7:486 (1968).

4. Goldman, R., Kedem, O. & Katchalski, E., Biochemistry,
    7:4518 (1968).

5. Goldman, R., Kedem, O. & Katchalski, E., Biochemistry
    10:165 (1971).

6. Broun, G., Avrameas, S., Selegny, E. & Thomas, D.,
    Biochem. Biophys. Acta 185:260 (1969).

7. Selegny, E., Avrameas, S., Broun, G. & Thomas, D.,
    C.R. Acad. Sci. C. 266:1931 (1968).

8. Karube, I. & Suzuki, S., Biochem. Biophys. Res. Comm.
    47:51 (1972).

9. Bareli, A. & Katchalski, E., J. Biol. Chem. 238:1690 (1963).

10. Lilly, M.D., Hornby, W.E. & Crook, E.M., Biochem. J. 100:718 (1966).

11. Thomas, D., Broun, G. & Selegny, E., Biochimie 54:229 (1972).

12. Broun, G., Thomas, D., Gellf, G., Domurado, D., Berjonneau, A.M. & Guillon, C., Biotechnol. Bioeng. 15: 359 (1973).

13. Broun, G., Selegny, E., Tran Minh, C. & Thomas, D., FEBS Ltrs 7:223 (1970).

14. Thomas, D., Tran Minh, C., Gellf, G., Domurado, D., Paillot, B., Jacobsen, R. & Broun, G., Biotech. Bioeng. Symp. 3:299 (1972).

15. Broun, G., Thomas, D. & Selegny, E., J. of Membrane Biol. 8:373 (1972).

16. Barbotin, J.N. & Thomas, D., J. Histochem. Cytochem. 22: 1048 (1974).

17. Thomas, D., Bourdillon, C., Broun, G. & Kernevez, J.P., Biochemistry 13:2995 (1974).

18. Naparstek, A., Romette, J.L., Kernevez, J.P. & Thomas, D., Nature 249:490 (1974).

19. Naparstek, A., Thomas, D. & Caplan, S.R., Biochem. Biophys. Acta 323:643 (1973).

20. Thomas, D., Golbeter, A. & Lefever, R., Biophysical Chemistry, in press.

21. Tosa, T., Mori, T. & Chibata, I., J. Ferment. Technol. 49:522 (1971).

22. Gellf, G., Thomas, D., Broun, G. & Kernevez, J.P., Biotechnol. Bioeng. 16:315 (1974).

23. Gellf, G. & Boudrant, J., Biochem. Biophys. Acta 334: 467 (1974).

24. Gellf, G., Henry, J., Kernevez, J.P. & Thomas, D.,
    C.R. Acad. Sci. D. 2265 (1973).

25. Berjonneau, A.M., Thomas, D. & Broun, G., Pathologie
    Biologie 22:497 (1974).

26. Paillot, B., Remy, M.H., Thomas, D. & Broun, G., Patho-
    logie Biologie 22:491 (1974).

MASS TRANSFER AND REACTION WITH THE MULTICOMPONENT MICRO-CAPSULES

Kozo Nakamura[*] and Yoshiro Mori[**]
Department of Agricultural Chemistry, Faculty
of Agriculture[*]; Department of Chemical Engineering, Faculty of Engineering[**]
University of Tokyo, Bunkyo-ku, Tokyo 113

Microencapsulation is the modern technique being used
in the fields related to printing, pharmaceutics, and food
to protect a material from its environment, modify its physical properties, mask its taste or odor control its re-
lease. It is particularly useful in the medical field,
since this technique will embody a concept of artificial
cell (1). In this paper the authors will make a literature
survey of the microencapsulation of enzyme, report their
experiment on the capsule size measurement, and analyze mass
transfer and reaction with the multicomponent microcapsules.

LITERATURE SURVEY OF THE MICROENCAPSULATION OF ENZYME

The survey of literature except patents shows that the
enzyme solution has been encapsulated by either of three
methods accompanying each characteristic step; (1) interfacial polymerication, (2) interfacial coaservation or (3)
drying in liquid. The features of those methods are summarized in Table 1.

There are several factors to be considered for evaluation of the encapsulation methods such as mechanical
strength of the semipermeable membrane, activity of the encapsulated enzyme, and its stability. The activity depends
not only on the performance of the encapsulation process,
but also on the conditions in the reactor containing the
microcapsules. Such proper analysis of the data as done by
Mogensen and Vieth (5) are more needed.

99

TABLE 1   SUMMARY OF LITERATURE SURVEY

Method	Enzyme (1)	W/O Emulsion						Ratio of Activity (8) [%]	Note	Ref.
		Aqueous Phase			Organic Phase		Membrane Formation			
		Reactant (2)	Additive (3)	Buffer	Solvent (4)	Emulsifier or Wall Material (5)	Solvent (6)	Reactant or Wall Material (7)			
1	Asparaginase (0.3-14.3 mg/ml) 1 ml	HD 1 ml	L-Asp.A Casein	0.45M NaHCO₃-Na₂CO₃ 1.5 ml	CF/CH (0-1.0) 20 ml	Span 85 (0.5-5.0v/v%)	CF/CH (0-1.0) 15 ml	SC	0-37		2
	Carbonic Anhydrase (12.5-37.5mg/ml) 0.2 ml	HD 0.4M in buffer	BSA (50 %) 0.9 ml	0.45M NaHCO₃-Na₂CO₃ 0.5 ml	CF/CH (0.25) 10 ml	Span 85 (1.5 %)	CF/CH (0.25) 10 ml	SC (0.09 %)	0.74-12	$K_m'=K_m$	3
	β-Galactosidase (1.3 mg/ml)	HD 0.4M in buffer 1.5 ml	BSA (50 w/v%) 1.5 ml	0.45M NaHCO₃ 15 ml	CF/CH (0.25) 15 ml	Span 85 (1.5 %)	CF/CH (0.25) 15 ml	TD (0.018 M)	21	$K_m'=K_m$	4
2	Catalase		Hb 1 gr.	Tris-buffer (pH 8.6) 2.0 ml Water 8.0 ml	Ether 100 ml		Ether 100 ml nBB 100 ml	CN (collodion) 100 ml	2.5-5.0 (9)		5
	Urease (10 mg/ml)	Water	"	"	"		"	"	— (9)	$K_m'=K_m$	
3	Catalase Hemolysate Lipase Urease (10-150 mg/ml)	Water 1-30 ml			Benzene 5-150 ml	PS,SD,EC (5-10 w/v%)	Water containing surface active agent or Gelatin 33-1000 ml		11 (10) (Catalase) 78-87 (crude Urease) 18 (pure Urease)	Effi. of MC 30-60%	6

(1) The concentration calculated equals the amount of enzyme added to a unit volume of the aqueous phase.

(2) HD=1,6-Hexanediamine (3) L-Asp.A=L-Asparatic acid, BSA=Bovine serum albumin, Hb=Hemoglobin (4) CF=Chloroform, CH=Cyclohexane

(5) PS=Polystyrene, SD=Silicone derivative, EC=Ethyl cellulose (6) nBB=n-Butyl benzoate

(7) SC=Sebacoyl chloride, TD=Terephthaloyl chloride, CN=Cellulose nitrate

(8) The ratio is defined as the activity of intact capsules devided by that of native enzyme.

(9) The catalase separated from the crushed capsules has only 5-10 % of its original activity, while the encapsulation has no effect on the urease activity.

(10) The catalase separated from the crushed capsules has 70 % of its original activity.

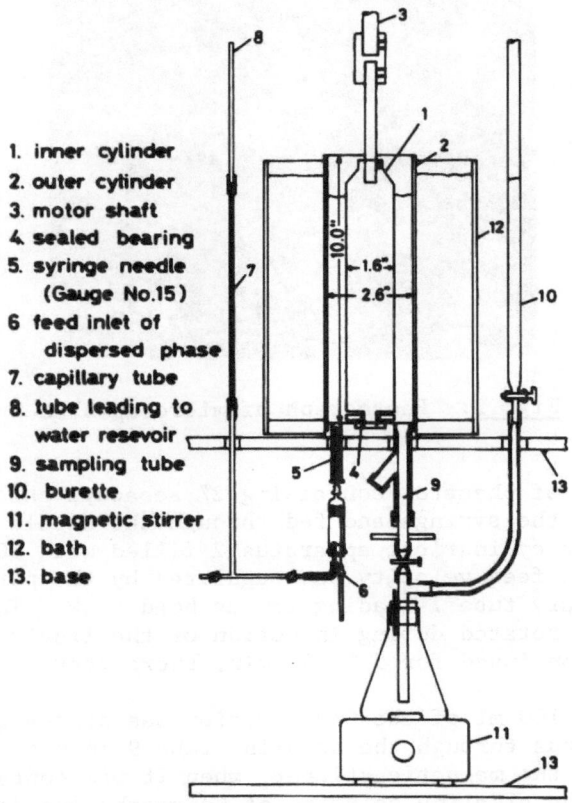

1. inner cylinder
2. outer cylinder
3. motor shaft
4. sealed bearing
5. syringe needle
   (Gauge No.15)
6 feed inlet of
   dispersed phase
7. capillary tube
8. tube leading to
   water resevoir
9. sampling tube
10. burette
11. magnetic stirrer
12. bath
13. base

<u>Fig. 1</u>: Experimental apparatus for microencapsulation

CAPSULE SIZE MEASUREMENT

The capsule size distribution should be known in advance when the kinetics of the encapsulated enzyme is analyzed. The experiment on the microencapsulation by interfacial polymerization was made in the apparatus shown in Fig. 1 to test its availability for the size measurement of liquid drops. It was not originally planned for immobilization of enzyme.

The liquid encapsulated, i.e. phenetol, was nonaqueous ether with the almost same specific gravity as that of water.

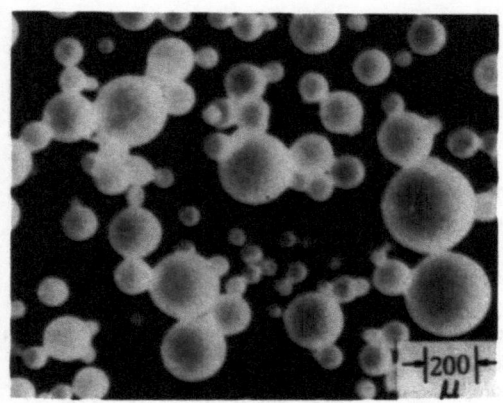

Fig. 2:   Photograph of microcapsules

3 to 10 ml of phenetol containing 2% sebacoyl chloride was
charged in the syringe and fed through the needle 5 into
the annular cylindrical apparatus 2 filled with 500 ml of
water.   The feed velocity was regulated by the pore size of
the capillary tube 7 leading to the head tank.   The inner
cylinder 1 rotated during injection of the liquid, and its
rotation continued for 5 to 30 min. thereafter.

About 100 ml of the O/W emulsion was discharged from
the apparatus through the sampling tube 9 into the flask
mounted on the magnetic stirrer, when it was contacted with
20 ml of the alkaline solution of hexamethylene diamine
flowing down from the burette 10.   The solution suspending
microcapsules was stirred for about 10 min., and 80 ml of
0.2% Tween 20 solution was poured into it for uniform dispersion of the microcapsules.   This dilute suspension was
stirred further for several minutes after it was neutrali-
zed with about 8.8 ml of 10% hydrochloric solution.

The size distribution of the microcapsules was directly
measured in the suspension by Coulter counter or on the photograph as shown in Fig. 2 by Zeiss size analyzer.   The
Sauter mean d_S, i.e. the capsule mean volume divided by its
mean surface area, was calculated with the size distribution
measured and plotted in Fig. 3.   It shows that the mean size
decreases linearly on the log-log graph paper with the speed
of rotation set at the formation of liquid drops.   The direct method gave the results agreeable with those by the

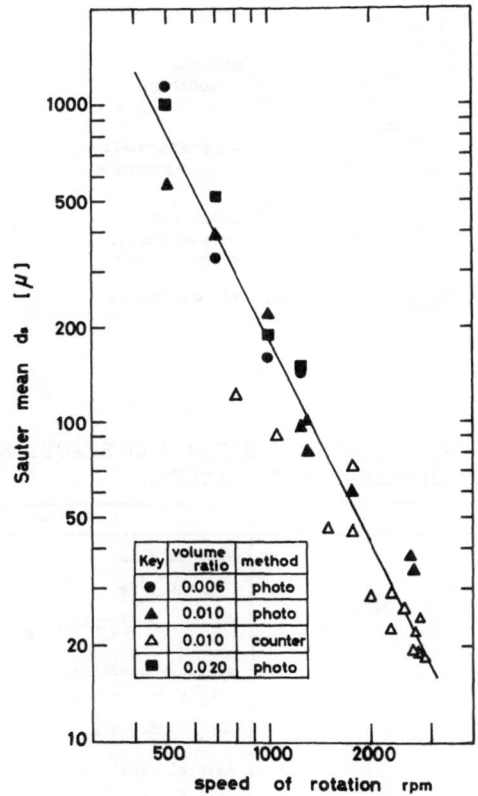

<u>Fig. 3</u>: Relationship between liquid drop size and speed of rotation.

photographic one, and proved convenient and useful for the capsule size measurement.

ANALYSIS ON THE KINETICS WITH THE MULTICOMPONENT MICRO-CAPSULES

Dr. Chang originally used the encapsulated urease and adsorbents for artificial kidney to decompose urea in the blood and to remove the toxic product ammonia (1). The other group studied also on this type of artificial kidney to process the dialyzing fluid (7). This section of the paper is assigned to analysis on mass transfer and reaction

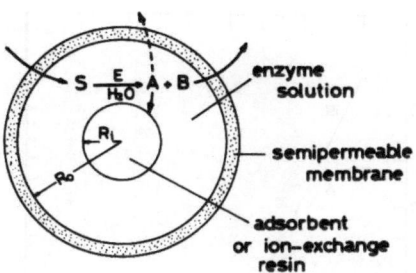

<u>Fig. 4</u>: Model of capsule.

TABLE 2 EQUATIONS OF MASS TRANSFER AND REACTION WITH THE MODEL CAPSULES-REACTOR SYSTEM

	Inside Capsule	In Bed or In Bath	
Substrate	**Mass balance:** $\frac{\partial y_1}{\partial t} = D_1 \left(\frac{\partial^2 y_1}{\partial r^2} + \frac{2}{r} \frac{\partial y_1}{\partial r} \right) - k_r y_1$ **Boundary condition:** $D_1 \left(\frac{\partial y_1}{\partial r} \right) = K_{1+m} (x_1 - y_1) \; ; \; r = R_o$ $D_1 \left(\frac{\partial y_1}{\partial r} \right) = 0 \; ; \; r = R_i$ **Initial condition:** $y_1 = 0 \; ; \; t = 0$	**Mass balance:** $\frac{\partial x}{\partial t} = E_z \frac{\partial x}{\partial z^2} - u \frac{\partial x}{\partial z}$ $\qquad - \frac{3(1-\alpha)}{\alpha R_o} K_{1+m} \left(x - (y)_{R_o} \right)$ **Boundary condition:** $-E_z \left(\frac{\partial x}{\partial z} \right) = u (x_f - x) \; ; \; z = 0$ $-E_z \left(\frac{\partial x}{\partial z} \right) = 0 \; ; \; z = L$ **Initial condition:** $x = 0 \; ; \; t = 0$	**Fixed Bed**
Product to be Adsorbed	**Mass balance:** (in solution) $\frac{\partial y_2}{\partial t} = D_2 \left(\frac{\partial^2 y_2}{\partial r^2} + \frac{2}{r} \frac{\partial y_2}{\partial r} \right) + k_r y_1$ (in adsorbent) $\frac{\partial c_2}{\partial t} = \frac{D_a}{\beta} \left(\frac{\partial^2 c_2}{\partial r^2} + \frac{2}{r} \frac{\partial c_2}{\partial r} \right) - k_a \left(c_2 - \frac{n_2}{K_a} \right)$ $\frac{\partial n_2}{\partial t} = k_a \left(c_2 - \frac{n_2}{K_a} \right)$ **Boundary condition:** $D_2 \left(\frac{\partial y_2}{\partial r} \right) = K'_{1+m} (x_2 - y_2) \; ; \; r = R_o$ $D_2 \left(\frac{\partial y_2}{\partial r} \right) = \frac{D_a}{\beta} \left(\frac{\partial c_2}{\partial r} \right) \; ; \; r = R_i$ $y_2 = c_2 \; ; \; r = R_i$ $\frac{D_a}{\beta} \left(\frac{\partial c_2}{\partial r} \right) = 0 \; ; \; r = 0$ **Initial condition:** $y_2 = 0, \; c_2 = 0, \; \text{and} \; n_2 = 0 \; ; \; t = 0$	**Mass balance:** $\frac{dx}{dt} = - \frac{3(1-\alpha)}{\alpha R_o} K_{1+m} \left(x - (y)_{R_o} \right)$ **Initial condition:** $x_1 = x_o \; \text{and} \; x_2 = 0 \; ; \; t = 0$	**Stirred Suspension**

with the model capsules as shown in Fig. 4.

The mass balance of the substrate and the produce makes the equations in Table 2, when the type of reaction is first order. The assumptions are used:

(1) The contraction and swelling of the capsule are neglected, and so is the cross effect in solute permeation.

(2) The rate of adsorption and the relationship of adsorption equilibrium are both linear.

The concentration distribution of the substrate is derived at the steady state, when the type of reactor is a fixed bed (8);

$$\frac{x}{x_f} = 2 \exp(\frac{PeB}{2} h) \frac{(1+\gamma) \exp[\gamma PeB (1-h)/2]-(1-\gamma) \exp[-\gamma PeB (1-h)/2]}{(1+\gamma)(1+\gamma) \exp(\gamma PeB/2)-(1-\gamma)(1-\gamma) \exp(-\gamma PeB/2)} \tag{1}$$

where $PeB = \frac{uL}{E_z}$, $N_r = [\frac{3(1-\alpha)}{\alpha R_o}](K_o L/u)$, $h = \frac{z}{L}$, and $\gamma = \sqrt{1+4N_r/PeB}$

The overall mass transfer coefficient can be expressed by the equation:

$$1/K_o = 1/k_1 + 1/mk_m + 3/[(1-\lambda^3) R_o k_r \eta] \tag{2}$$

The effectiveness factor numerically calculated by Hovarth and Engasser (9) is applicable to η in Eq. 2, since their pellicular heterogeneous catalysts have the same shell configuration as the model capsule considered in this paper. The effectiveness factor, however, can be calculated analytically, if the order of reaction is zero or first;

for zero order reaction,

$$\eta = 1 \quad \text{if} \quad \phi < \frac{1}{1-\lambda} \sqrt{6/(1+2\lambda)} \tag{3}$$

$$\eta = \frac{1-\lambda_c^3}{1-\lambda^3} \quad \text{if} \quad \phi > \frac{1}{1-\lambda} \sqrt{6/(1+2\lambda)} \tag{4}$$

for first order reaction,

$$\eta = \frac{3}{(1-\lambda^3)} [\frac{1 + (\lambda\phi) \tanh(\phi(1-\lambda))}{\lambda\phi + \tanh(\phi(1-\lambda))} - \frac{1}{\phi}] \tag{5}$$

where the diffusional-kinetic modulus ϕ is $k_r R_o/D y_o$ for zero order reaction and $k_r R_o/D$ for first order reaction. λ_c in Eq. 4 is the dimensionless extinction radius given by the equation:

$$\phi^2 (1-\lambda_c)^2 (1+2\lambda_c) - 6 = 0 \qquad (6)$$

Fig. 5 is the graphical representation of Eqs. 3, 4, and 5 showing how the effectiveness factor increases with the ratio of radius λ at the constant value of ϕ. The capsule has larger effectiveness factor as expected when its shell part becomes thinner. The superficial activity pertaining to a single capsule depends, however, not only on the effectiveness factor but also on the volume of the active part. The product of η and $1-\lambda^3$ is calculated and normalized by that of a spherical capsule. Fig. 6 shows that the superficial activity is always larger for the capsule with a thicker active part, which can be also expected.

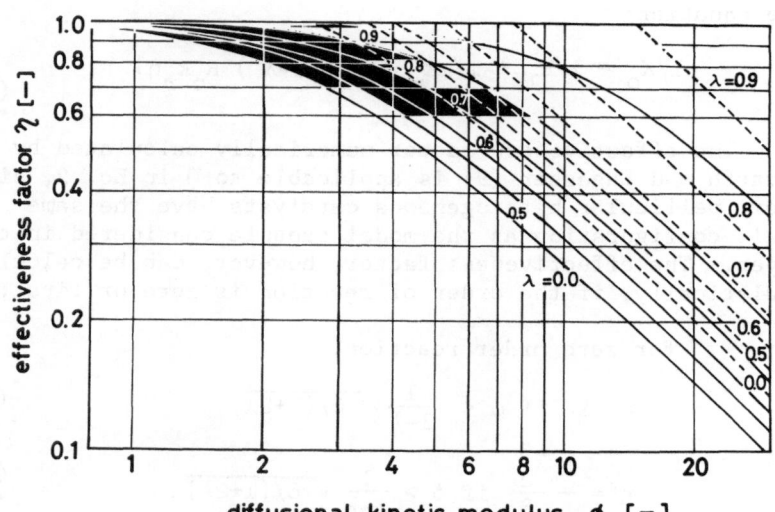

Fig. 5: Effectiveness factor for zero and first order irreversible reaction in the model capsule of Fig. 4 (dotted lines : zero order and solid lines : first order).

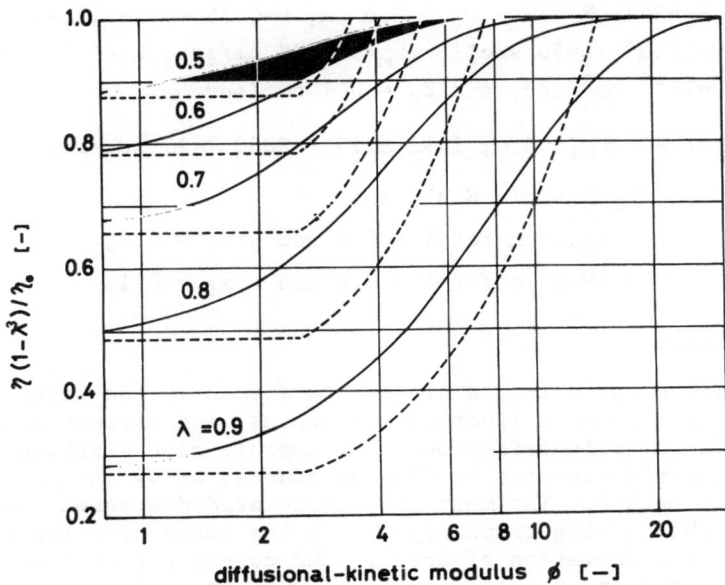

<u>Fig. 6</u>: Rate of reaction in the shell-structured capsule
normalized by that in the spherical one (dotted
lines : zero order and solid lines : first order).

The diffusional-kinetic modulus ϕ in this paper is different in definition from that used by Hovarth and Engasser.
It was assured, however, that Eq. 3 gives the value η
agreeable with their numerically calculated one. On the
other hand it was not possible precisely to compare the
effectiveness factor for zero order reaction with theirs,
since Hovarth and Engasser did not consider the substrate
exhaustion phenomena included in Eqs. 3 and 4.

The results of this analysis are useful for the experimental data to be interpreted properly, and the membrane permeability can be calculated from the measurable
mass transfer coefficient K_0. The relationship of Eq. 1
gives the column length to be needed for a given conversion
of the substrate. An example of calculation shows the possibility of a compact artificial kidney with the encapsulated enzyme as reported by Levine and LaCourse (10).

Data: $R_o=25\,\mu$, $R_i=0\,\mu$, $D_t = 2.0$ cm, $u=0.35$ cm/sec, $x_f=0.004$ g-mol/cm^3, $k_r=50$ sec^{-1}, $mk_m=5 \cdot 10^{-4}$ cm/sec, $k_1=6 \cdot 10^{-2}$ cm/sec, $D=10^{-5}$ cm^2/sec, and $E_z = 0.1$ cm^2/sec.

Results: $\phi \quad 6$, $\eta \simeq 0.4$, $1/K_o \simeq 15 + 2000 + 17$ sec/cm,

$\quad\quad\quad x/x_f \simeq \exp(-N_r h)$, and

$\quad\quad\quad L = 11.0$ cm for $\alpha = 0.6$ and $x/x_f = 0.1$,

$\quad\quad\quad L = 19.6$ cm for $\alpha = 0.9$ and $x/x_f = 0.1$

CONCLUSION

The organic liquid drops were formed in the water fil-ling the annular cylindrical apparatus, and encapsulated by interfacial polymerization. The capsule size distribution was directly measured by Coulter counter or by the photo-graphic method. The mean size calculated decreased linear-ly on the log-log graph paper with the speed of rotation set at the formation of drops. The convenient electronic method gave the results agreeable with those by the photo-graphic method.

The basic equations of mass transfer and reaction were expressed for the multicomponent microcapsules-reactor sys-tem, and the effectiveness factor was analytically derived for the substrate undergoing zero or first order reaction. An example of calculation resulted in the conclusion as re-ported by Levine and LaCourse that the artificial kidney could be compact, if the encapsulated urease is used.

ACKNOWLEDGEMENT

The capsule size measurements were done by K.N. while at the National Research Council of Canada as a Post Doc-torate Fellow. The authors thank N.R.C.C. and Dr. C.E. Capes for permitting publication of the data.

NOMENCLATURE

c: solute concentration in the pore of adsorbent (g-mole/cm^3)
D and D_a: interparticle and intraparticle diffusivity of solute (cm^2/sec).
D_t: column diameter (cm).
K_a: adsorption equilibrium constant (-).

K_{1+m}: surface mass transfer coefficient (cm/sec).
K_o: overall mass transfer coefficient (cm/sec).
k_a: adsorption rate constant (sec^{-1})
k_r: rate constant of zero or first order reaction
  (g-mole/cm^{3}sec) or (sec^{-1}).
L: axial column length (cm).
m: partition coefficient (-).
n: concentration of the solute adsorbed on a unit volume
  of the adsorbent (g-mole/cm^3).
r: radial coordinate (cm).
R_o and R_i: radius of the capsule and of the adsorbent (cm).
t: time (sec).
u: liquid velocity in the column (cm/sec).
x and y: solute concentration in the bulk liquid and in
  the encapsulated liquid (g-mole/cm^3).
y_o: y at r = R (g-mole/cm^3).
z: axial coordinate (cm).

α : void fraction in the bed (-).
β : void fraction in the adsorbent (-).
η: effectiveness factor (-).
λ : ratio of radius (= R_i/Ro) (-).
λ_c: extinction radius (-).
ϕ: kinetic-diffusional modulus (-).

REFERENCES

1. Chang, T.M.S., "Artificial Cells", Charles C. Thomas
   (1972).

2. Mori, T., Saito, T., Matsuo, Y., Tosa, T. and Chibata,
   I., Biotech. & Bioeng., 14: 663 (1972).

3. Boguslaski, R.C. and Janik, A.M., Biochimica et Biophy-
   sica Acta 250: 266 (1971).

4. Ostergaard, J.C.W. and Martiny, S.C. Biotech. & Bioeng.
   15: 561 (1973).

5. Mogensen, A.O. and Vieth, W.R., Biotech. & Bioeng. 15:
   467 (1973).

6. Kitajima, M., Miyano, S. and Kondo, A. Kogyo Kagaku
   Zashi 72: 493 (1973).

7. Sparks, R.E., Salemme, R.M., Meier, P.M., Litt, M.H. and
   Linden, O., Trans. Amer. Soc. Artif. Int. Organs 15: 353
   (1969).

8. Danckwerts, P.V., Chem. Eng. Sci., 2: 1 (1953).
   Yagi, Y. and Miyauchi, T. Kagaku Kogaku (Chem. Eng.,
   Japan) 19: 507 (1955).

9. Hovarth, C. and Engasser, J., Ind. Eng. Chem. Fundam.,
   12: 229, (1973).

10. Levine, S.N. and LaCourse, W.C., J. Biomed. Mat. Res.
    1: 275 (1967).

APPLICATIONS OF IMMOBILIZED ENZYMES AND IMMOBILIZED MICROBIAL CELLS FOR L-AMINO ACID PRODUCTION

Ichiro Chibata, Tetsuya Tosa, Tadashi Sato,
Takao Mori and Kozo Yamamoto
Research Laboratory of Applied Biochemistry
Tanabe Seiyaku Co., Ltd., 16-89, Kashima-3-chome
Yodogawa-ku, Osaka, Japan

Recently, utilization of L-amino acids for medicines, food and animal feed has been rapidly developing, and the economical production of optically active amino acids and the related compounds has been needed.

Over the past 10 years, we have been studying on industrial application of immobilized enzymes and immobilized microbial cells for production of L-amino acids and the related compounds. The detailed papers on these studies have been published (1-16) or will be published in near future (17). Thus, our studies are reviewed in this presentation.

I. PRODUCTION OF L-AMINO ACIDS BY IMMOBILIZED AMINO-ACYLASE

At present, fermentative and organic synthetic methods are employed for the industrial production of L-amino acids instead of conventional isolation method from protein hydrolysate. Amino acids produced by the fermentative method are L-form, whereas amino acids produced by chemical synthetic method are optically inactive racemic mixture of L- and D-isomers. To obtain the L-amino acid from the chemically synthesized mixture of DL-form, optical resolution is necessary. Optical resolution of racemic amino acids can be carried out by several methods. Among these methods, the enzymatic method using mold aminoacylase is one of the advantageous procedures for the industrial production of optically pure L-amino acids. The reaction catalyzed by the enzyme is as follows.

111

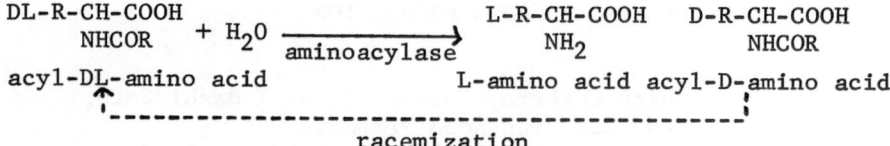

racemization

That is, chemically synthesized acyl-DL-amino acid is
asymmetrically hydrolyzed by aminoacylase, and L-amino acid
and unhydrolyzed acyl-D-amino acid are produced in the reaction mixture. After being concentrated, both compounds
are easily separated by the difference in their solubilities.
Acyl-D-amino acid is racemized, and reused for the resolution procedure. This mold aminoacylase has versatile substrate specificity and many kinds of L-amino acids can be
produced. From 1954 to 1969, this enzymatic resolution
method had been employed in our company, Tanabe Seiyaku Co.,
Ltd., for the industrial production of several L-amino acids.

However, the enzyme reaction had been carried out in
batch process by incubating a mixture containing substrate
and soluble enzyme. Thus, the procedure had some disadvantages for industrial purpose. To overcome these disadvantages, the immobilization of aminoacylase and the continuous optical resolution of DL-amino acids using columns
packed with the immobilized enzyme has been studied since
1960 in our Research Laboratory. As a result, the efficient
and automatically controlled enzyme reaction system was ac-
complished, and since 1969 we have been industrially operating several series of the enzyme reactors in our plants for
the production of L-amino acids such as L-methionine, L-
phenylalanine, L-tryptophan, L-valine, etc. This new procedure gives us satisfactory results, and this is considered
to be the first industrial application of immobilized enzyme
in the world.

1.) Preparation and Selection of Immobilized Amino-
acylase Suitable for Industrial Purpose

On the preparation of immobilized enzymes, a number of
methods have been known, and we tested these many methods
for immobilization of aminoacylase to obtain the most suitable enzyme preparation for industrial purpose (1,2,4,9,10,
11).

As the results, relatively active and stable immobilized aminoacylases were obtained by ionic binding to DEAE-Sephadex (4), covalent binding to iodoacetyl cellulose (9) and entrapping into polyacrylamide gel lattices (10). Thus, the enzymatic properties of these 3 kinds of immobilized aminoacylases were studied and compared with those of the native one to select the most suitable preparation (6,9,10, 11). On the heat stability of immobilized aminoacylases, the DEAE-Sephadex-aminoacylase showed the highest stability among these 3 immobilized preparations and native enzyme (10). Further, it showed strong resistance towards proteases, organic solvents, and protein denaturing agents (6,7). For the industrial application of immobilized aminoacylase, it is necessary to satisfy a number of conditions. Characteristics of immobilized aminoacylases are summarized in Table I.

From the results shown in the table, we chose immobilized DEAE-Sephadex-aminoacylase as one of the most advantageous enzyme preparation for the industrial production of L-amino acids, because 1.) preparation is easy, 2.) cost of immobilization is low, 3.) operation stability is high, and 4.) regeneration of deteriorated immobilized aminoacylase is possible.

2.) Continuous Production of L-Amino Acids

When we started the studies on the industrial application, very little was known about chemical engineering of enzyme columns packed with an immobilized enzyme as a solid catalysts, and no reports had been published on the industrial application. Thus, to find out most suitable conditions for continuous production of L-amino acids, effects of pH of substrate, temperature and flow rate, stability, regeneration, kinetics, pressure drop and operation cost of column were investigated (4-6,8,11). From the results obtained, the most efficient and automatically controlled enzyme reaction system was designed as shown in Figure 1, and since the summer of 1969 we have been operating this enzyme reactor system. As the results, the average overall production cost of the amino acids by immobilized aminoacylase process is about 60% of that of the conventional bath process using the soluble enzyme due to remarkable reduction of amounts of substrate and aminoacylase, and labor cost.

TABLE I

Characteristics of Typical Immobilized Aminoacylases

Characteristics	Immobilized aminoacylases		
	Ionic binding to DEAE-Sephadex	Covalent binding to iodeacetylcellulose	Entrapping by polyacrylamide
Preparation	easy	difficult	medium
Cost of immobilization*	low	high	moderate
Binding force	medium	strong	strong
Operation stability (half-life**, days)	65 (50°C)	—	48 (37°C)
Regeneration***	possible	impossible	impossible

* Compared from the basis for unit production of L-amino acids.

** The time required for 50% of the enzyme activity to be lost.

*** Regeneration of deteriorated immobilized aminoacylase column after operation for long period.

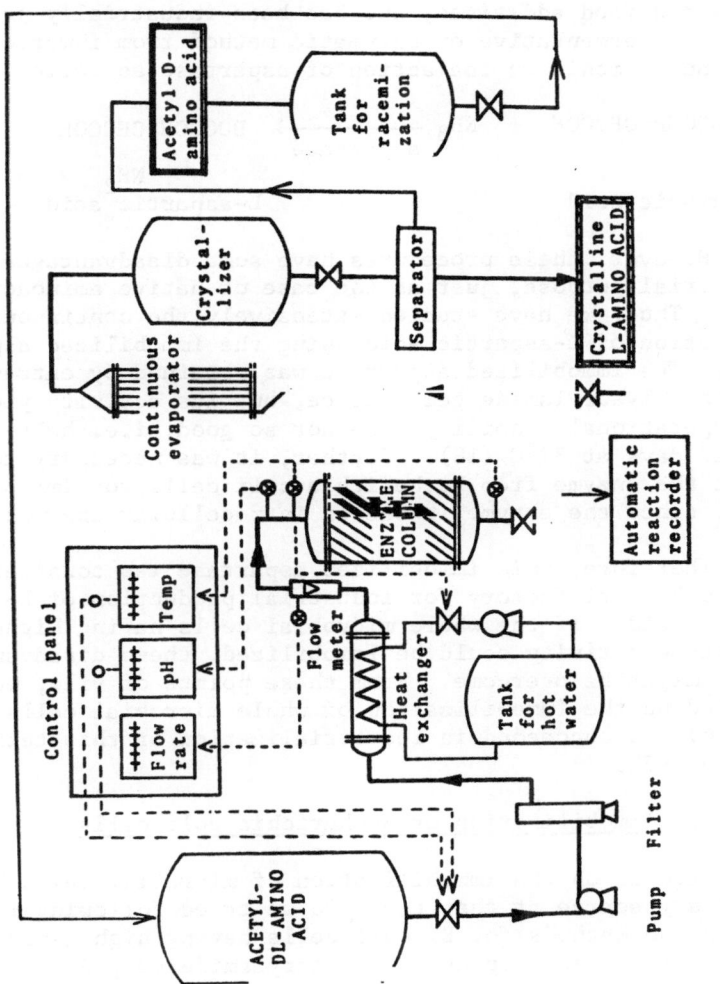

Fig. 1: Flow Diagram for Continuous Production of L-Amino Acid by Immobilized Aminoacylase

II. PRODUCTION OF L-ASPARTIC ACID BY IMMOBILIZED ESCHERICHIA COLI CELLS

Succeeding to the industrialization of continuous optical resolution using immobilized aminoacylase, we investigated the continuous production of L-aspartic acid by immobilized aspartase (18). The acid is widely used as medicines and food additives, and has been industrially produced by fermentative or enzymatic method from fumaric acid and ammonia by the action of aspartase as follows.

$$HOOCCH=CHCOOH \; + \; NH_3 \xrightarrow[\text{aspartase}]{} HOOCCH_2CHCOOH$$

$$\underset{\text{fumaric acid}}{} \qquad\qquad \underset{\text{L-aspartic acid}}{\overset{NH_2}{|}}$$

However, these procedures have some disadvantages for industrial purpose, just as the case of native aminoacylase. Thus, we have studied extensively the continuous production of L-aspartic acid using the immobilized aspartase. The immobilized aspartase was obtained by entrapment into a polyacrylamide gel lattice, but its activity yield and operational stability were not so good, i.e. half-life was 27 days at $37^{\circ}C$ (18). Further, it was necessary to extract the enzyme from Escherichia coli cells for immobilization, since the enzyme is one of intracellular enzymes.

Therefore, this immobilized aspartase was considered not to be satisfactory for industrial production of L-aspartic acid. If the whole microbial cells having higher aspartase activity could be immobilized, these disadvantages might be overcome. From these points of view, we studied on the immobilization of whole microbial cells (12,14) and succeeded in industrialization of this technique in 1973.

1.) Immobilization of Escherichia coli cells

Studies on the immobilization of microbial cells had been very scarce at that time, so we tried following immobilization methods for E. coli cells having high aspartase activity; 1.) entrapping by polyacrylamide gel, 2.) crosslinking by glutaraldehyde or 2,4-toluene diisocyanate, and 3.) encapsulation by polyurea produced from 2,4-toluene diisocyanate and hexamethylene diamide. Among these methods the most active immobilized E. coli cells were obtained by entrapping the cells into polyacrylamide gel-lattice.

To prepare the most efficient immobilized microbial cells by this polyacrylamide gel method, the type and concentration of bifunctional reagents and the concentration of acrylamide monomer were investigated. As the results, following optimum conditions for immobilization were decided.

E. coli cells (10 kg, wet weight) collected from cultured broth are suspended in 40 liters of physiological saline. To this suspension are added 7.5 kg of acrylamide, 0.4 kg of N,N'-methylenebisacrylamide, 5 liters of 5% β-dimethylaminopropionitrile, and 5 liters of 2.5% potassium persulfate. The mixture is allowed to stand at below 40°C for 10~15 minutes and the resulting stiff gel is made 2 3 mm granules. The aspartase activity of immobilized E. coli obtained under the optimum conditions is 12,000~16,000 μ mole/hr/gram of wet cells.

An interesting phenomenon was observed with these cells. When the immobilized E. coli cells were suspended at 37°C for 24~48 hours in substrate solution, the activity increased 9~10 times. This phenomenon was also recognized when intact cells were incubated in the same solution. This activation was considered to be either adaptive formation of aspartase-protein in the presence of substrate or increase of membrane permeability for substrate and/or product due to autolysis of E. coli cells in the gel lattice.

Thus, in order to investigate the adaptive formation of the enzyme, fresh cells or fresh immobilized cells were incubated in 1 M substrate solution for 48 hours at 37°C in the absence or presence of chloroamphenicol at the concentrations completely inhibiting protein synthesis. The results indicated that the enzyme activities increased, even in the presence of chloroamphenicol. Therefore, this activation is considered not to be the result of protein synthesis but due to the increased permeability by autolysis of E. coli cells in the gel lattice. This was also confirmed by the electron micrograph of immobilized E. coli cells after activation, indicating lysis of cells occurred. Of course, even lysis of the cells occur, the aspartase does not leak out from the gel lattice, though the substrate and the product easily pass through the gel lattice.

2.) Continuous Production of L-Aspartic Acid

To find out the most suitable conditions for continuous production of L-aspartic acid from ammonium fumarate, the enzymatic properties of immobilized E. coli cells were investigated (12).

The immobilized cells showed an optimal activity at pH 8.5 (same as native aspartase), whereas the optimal pH of the intact cells is 10.5. The effect of temperature on the formation of L-aspartic acid by the immobilized cells was compared with that of the intact cells, and the optimal temperature was found to be 50°C in both preparations.

On the effects of metal ions, although the native and immobilized aspartases are activated by Mn^{++}, the formation of L-aspartic acid by intact and immobilized cells was not accelerated by this metal ion. On the other hand, investigation of the protective effects of various metal ions against heat inactivation of intact and immobilized cells shoed that bivalent metal ions such as Ba^{++}, Ca^{++}, Mg^{++}, Mn^{++} and Sr^{++} protect the intact and immobilized cells. Further, these protective or stabilizing effects of the bivalent metal ions were investigated in the case of continuous formation of L-aspartic acid by the column process. As shown in Table II, Ca^{++}, Mg^{++} and Mn^{++} have stabilizing effect to aspartase activity of the immobilized E. coli during the continuous enzyme reaction.

Conditions for continuous production of L-aspartic acid from ammonium fumarate were investigated in detail by passing 1 M ammonium fumarate containing 1 mM Mg^{++}, pH 8.5, into a column packed with immobilized E. coli cells (13,14).

The maximal flow rate of substrate solution enabling the complete conversion of ammonium fumarate to L-aspartic acid was space velocity=0.8 at 37°C. The stability of the immobilized cell column was investigated by continuously passing a substrate solution for long period at various temperature as shown in Figure 2. The results showed that the deterioration of the activity depends on temperature and the immobilized cell column is very stable, and the half-life of the column was estimated to be 120 days at 37°C as shown in Table II.

TABLE II

Operation Stability of Immobilized Microbial Cells

Microorganisms	Enzymes	Addition of metals	Operation stability Half-life*, days	References
Escherichia coli	aspartase	Ca^{++}, Mg^{++}, Mn^{++}	120 (37°C)	13, 14
Pseudomonas putida	L-arginine deiminase	—	140 (37°C)	14, 15
Achromobacter liquidum	L-histidine ammonia-lyase	Ca^{++}, Co^{++}, Mg^{++}, Zn^{++}	180 (37°C)	14, 16
Escherichia coli	penicillin amidase	—	17 (40°C) 42 (30 C)	17

* The time required for 50% of the enzyme activity to be lost.

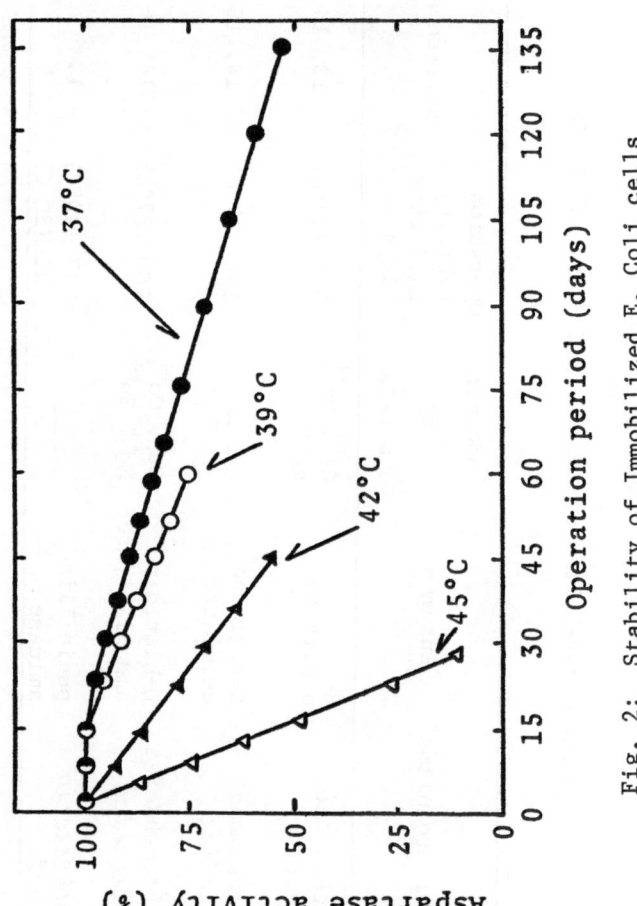

Fig. 2: Stability of Immobilized E. Coli cells Column at Various Temperature

According to these informations the aspartase reactor
system using immobilized E. coli cells was designed, which
is essentially same as the case of the immobilized aminoacylase system shown in Figure 1. Since the autumn of 1973
we have been industrially operating this new system. As
the results, the overall production cost of the immobilized
cell system is reduced to about 60% of the conventional
bath process using intact cells.

This is considered to be the first industrial application of immobilized microbial cells in the world.

III. OTHER APPLICATIONS OF IMMOBILIZED MICROBIAL CELLS

In succession to the continuous production of L-aspartic acid, we studied continuous methods for the efficient
production of useful organic compounds such as L-citrulline
(15), urocanic acid (16) and 6-aminopenicillanic acid (17)
by using immobilized microbial cells. As satisfactory results for production of these compounds have been obtained,
the outlines of their studies are presented in this chapter.

1.) Production of L-Citrulline by Immobilized Pseudo-   monas putida Cells

L-Citrulline is used for medicines, and has been produced from L-arginine by the action of microbial L-arginine
deiminase as follows.

$$H_2NCNHCH_2CH_2CH_2CHCOOH + H_2O \xrightarrow{\text{L-arginine deiminase}}$$
$$\underset{NH}{\|} \qquad \underset{NH_2}{|}$$

$$H_2NCONHCH_2CH_2CH_2CHCOOH + NH_3$$
$$\underset{NH_2}{|}$$

L-citrulline

In order to produce L-citrulline more advantageously,
we (15)studied the immobilization of Pseudomonas putida
ATCC 4539 having high L-arginine deiminase activity, and
succeeded in its immobilization by the polyacrylamide gel
method as in the case of E. coli cells.

The conditions for the continuous production of L-citrulline by using a column packed with the immobilized P. putida cells are as follows. When an aqueous solution of 0.5 M L-arginine hydrochloride (pH 6.0) is passed through the column at 37°C at flow rates below space velocity=0.26, the reaction is completed. From the column effluent, pure L-citrulline is obtained in a high yield by the concentration and the ion exchange resin treatments. As shown in Table II, the stability of the immobilized cell column is very high.

This technique is considered to be more advantageous for the production of L-citrulline than the batch method using microbial broth.

2.) Production of Urocanic Acid by Immobilized Achromobacter liquidum Cells

Urocanic acid is used as a sun-screening agent in the pharmaceutical and cosmetic fields, and is produced from L-histidine by the action of microbial L-histidine ammonia-lyase as follows.

$$HC=C-CH_2CHCOOH \xrightarrow[\text{ammonia-lyase}]{\text{L-histidine}} HC=CH-CH-CHCOOH + NH_3$$

L-histidine   Urocanic acid

We (16) tested the immobilization of several micro-organisms having the enzyme activity by polyacrylamide gel method, and compared the enzyme activities of the intact and immobilized cells. As the results, we chose Achromobacter liquidum IAM 1667, because the organism showed the highest activity after immobilization. Although the organism has urocanase activity which converts urocanic acid to imidazolone propionic acid, this activity was removed by a simple heat treatment (70°C, 30 minutes) before immobilization of the cells.

By using a column packed with the immobilized A. liquidum cells, the conditions for the continuous production of urocanic acid were investigated. When an aqueous solution of 0.25 M L-histidine (pH 9.0) containing 1 mM Mg^{++} was passed through the column at flow rates below space velocity=0.06, L-histidine was completely converted to urocanic acid. From the column effluent, pure urocanic acid is crystallized in a high yield by simple pH treatment. The enzyme activity of the column is very stable in the presence of Mg^{++} as shown in Table II.

This system is also more efficient for the industrial production of urocanic acid than the batch system using extracted enzyme or microbial broth.

3.) Production of 6-Aminopenicillanic Acid by Immobilized Escherichia coli Cells

Immobilized microbial cell technique was also applied for the production of 6-aminopenicillanic acid (6-APA) from penicillins (17).

6-APA is used as a starting material for the synthetic penicillins, and is industrially produced from penicillins by microbial penicillin amidase as follows.

$$\text{penicillin} + H_2O \xrightarrow[\text{amidase}]{\text{penicillin}} \text{6-APA} + RCOOH$$

However, these procedures have been carried out in a batch process by incubating a mixture of penicillins and microbial broth or enzyme extracted from microbial cells, therefore, they have some disadvantages for industrial purpose. To overcome these disadvantages, many studies on the continuous production of 6-APA by using a column packed with immobilized penicillin amidase have been published. However, in this case, it was necessary to extract the enzyme from microbial cells before the immobilization process, and in general the extracted penicillin amidase was unstable.

Thus, in order to develop a more efficient method, we have investigated the continuous production of 6-APA by the immobilized E. coli ATCC 9637 cells having high penicillin amidase activity.

E. coli cells used in this study have weakly penicillinase activity. However, in this E. coli the reaction
rate of penicillin amidase is much higher than penicillinase
which decomposes both penicilline and 6-APA. Therefore,
6-APA can be obtained by choosing appropriate conditions
without removing the penicillinase activity. That is, when
a solution of 0.05 M penicillin G in 0.01 M borate-phosphate
buffer (pH 8.5) was passed through the column packed with
the immobilized E. coli cells by polyacrylamide gel method
at flow rates of space velocity=0.12-0.24, 6-APA was efficiently formed, and isolated from the column effluent in
around 80% yield based on the used penicillin G.

As the immobilized cells column was relatively stable
as shown in Table II, the cost of enzyme-catalyst will be
thereby reduced. Further, since the production process of
6-APA can be automatically controlled, it is expected that
the labor cost is reduced. Therefore, this procedure is
considered to be advantageous than the continuous method
using immobilized penicillin amidase.

IV. DISCUSSION AND CONCLUSION

As described previously, several kinds of microbial
cells having an enzyme in high activity can be easily immobilized and stabilized by the entrapping method using
polyacrylamide gel. The continuous enzyme reaction by the
immobilized microbial cells will be employed advantageously
in the following cases. That is, 1.) when enzymes are intracellular, 2.) when enzyme extracted from microbial cells
are unstable, 3.) when enzymes are unstable during and after immobilization, 4.) when microorganism contains no other
enzymes which catalyze interfering side reaction, or those
interfering enzyme are readily inactivated or removed, and
5.) when substrates and products are not high molecular
compounds and can easily pass through the gel lattice.
Further, volume of fermentation broth for the unit production of desired compound is much smaller in the case of
continuous method using immobilized cells than in the case
of conventional fermentative method, indicating that the
former method is very advantageous in the point of water-
pollution in plant.

Therefore, in future, the studies on the immobilized
microbial cells will be developed extensively as well as

TABLE III

Factors to be Considered for Industrial Application of Immobilized Enzymes Including Immobilized Cells

Factors		Soluble enzyme	Immobilized enzymes	
		Batch system	Batch system	Column system
Cost	High / Low	suitable	suitable	suitable
Enzyme	Reuse	impossible	possible	possible
	Stability	low	moderate-high	high
Enzyme reaction Rate	Control High	difficult	difficult	easy
	Rate Low	suitable	suitable	suitable
Product	Purity	low	high	high
	Yield	low	high	high
Equipment	Initial cost	low	moderate	high
	Automation	difficult	difficult	easy
	Applicability	high	high	moderate
Running cost(Labor cost)		high	moderate	low
Scale merit		low	low	high

the immobilized enzymes.

According to our experiences for industrial applications of immobilized enzymes and immobilized microbial cells, followings are important factors.

1.) Cost of carriers or reagents for immobilization of enzymes or microbial cells.
2.) Activity of immobilized enzyme and yield from native enzyme or intact cells.
3.) Stability of immobilized enzyme or immobilized cells during operation.
4.) Regenerability of the deteriorated immobilized enzyme or microbial cells after long period operation.

Besides these conditions, a number of factors should be considered for industrial application of immobilized enzymes and immobilized cells as shown in Table III.

For further development of immobilized enzymes and immobilized cells, we hope, as a biochemist and enzymologist, to cooperate and collaborate with wide variety of scientists in catalytic chemistry, organic chemistry, polymer chemistry and chemical engineering.

REFERENCES

1. T. Tosa, T. Mori, N. Fuse, and I. Chibata; Enzymologia
    31, 214 (1966).

2. T. Tosa, T. Mori, N. Fuse, and I. Chibata; Enzymologia
    31, 225 (1966).

3. T. Tosa, T. Mori, N. Fuse, and I. Chibata; Enzymologia
    32, 153 (1967).

4. T. Tosa, T. Mori, N. Fuse, and I. Chibata; Biotech. Bioeng. 9, 603 (1967).

5. T. Tosa, T. Mori, N. Fuse, and I. Chibata; Agr. Biol.
    Chem. 33, 1047 (1969).

6. T. Tosa, T. Mori, and I. Chibata; Agr. Biol. Chem. 33,
    1053 (1969).

7. T. Tosa, T. Mori, and I. Chibata; Enzymologia 40, 49
    (1971).

8. T. Tosa, T. Mori, and I. Chibata; J. Ferment. Technol.
    49, 522 (1971).

9. T. Sato, T. Mori, T. Tosa, and I. Chibata; Arch. Bio-
    chem. Biophys. 147, 788 (1971).

10. T. Mori, T. Sato, T. Tosa, and I. Chibata; Enzymologia
    43, 213 (1972).

11. I. Chibata, T. Tosa, T. Sato, T. Mori, and Y. Matuo;
    Proceeding of IV International Fermentation Symposium:
    Fermentation Technology Today, p. 383 (1972).

12. I. Chibata, T. Tosa, and T. Sato; Appl. Microbiol. 27,
    878 (1974).

13. T. Tosa, T. Sato, T. Mori, and I. Chibata; Appl. Micro-
    biol. 27, 886 (1974).

14. I. Chibata, T. Tosa, T. Sato, T. Mori, and K. Yamamoto;
    Enzyme Engineering (ed. E.K. Pye and L.B. Wingard, Jr.,
    Plenum Press) Vol. 2, p. 303 (1973).

15. K. Yamamoto, T. Sato, T. Tosa, and I. Chibata; Biotech.
    Bioeng., 16, 1589 (1974).

16. K. Yamamoto, T. Sato, T. Tosa, and I. Chibata; Biotech.
    Bioeng. 16, 1601 (1974).

17. T. Sato, T. Tosa, and I. Chibata; Appl. Microbiol.,
    under contribution.

18. T. Tosa, T. Sato, T. Mori, Y. Matuo, and I. Chibata;
    Biotech. Bioeng. 15, 69 (1973).

KINETICS AND MASS TRANSFER CHARACTERISTICS OF GLUCOSE ISOMERASE IMMOBILIZED ON POROUS GLASS

Y.Y. Lee, Chemical Engineering Department, Auburn University, Auburn, Alabama;
K. Wun, Proctor & Gamble Co., Cincinnati, Ohio;
G.T. Tsao, School of Chemical Engineering, Purdue University, West Lafayette, Indiana

I. INTRODUCTION

Enzymic isomerization of glucose into fructose, in the past few years, has received considerable attention for its commercial importance in manufacturing sugar substitutes. A major limitation of enzymic glucose isomerization has been the relatively high cost of the enzyme. Immobilization of enzymes by covalent bonding to inert support has been the subject much research and appear to be well suited for use with glucose isomerase.

In this study, the glucose isomerase covalently attached to porous glass was investigated for important engineering factors, including intrinsic kinetics and mass transfer effects. Due to its reversible nature, the time-course kinetics of this reaction is more complicated than for simple, irreversible enzymic reactions. A generalized Michaelis-Menten equation for a reversible enzymic reaction was derived and its applicability to glucose isomerization was tested.

With the porous enzyme support, the diffusional interferences are created by two transfer processes: (a) transfer through external liquid film; and (b) transfer through porous enzyme support. These diffusional effects were experimentally verified for the current heterogeneous reaction system and analyzed by established theories. The effectiveness factor concept extended to film diffusion was introduced.

129

II. THEORY OF REVERSIBLE ENZYMIC REACTION

The published kinetics study of glucose isomerization has been limited to the initial stage of reaction wherein the uninhibited, conventional Michaelis-Menten kinetics is still applicable.

For the reverse glucose $\rightleftharpoons$ fructose reaction, following simple reaction mechanism will be assumed for derivation of the reaction rate equation.

$$S + E \underset{k_{-1}}{\overset{k_1}{\rightleftharpoons}} SE \underset{k_{-2}}{\overset{k_2}{\rightleftharpoons}} E + P \tag{1}$$

Under steady state assumption, the above mechanism leads to the following rate equation for the forward reaction.

$$v = \frac{EE_o Ck_1 k_2 S - k_{-1} k_{-2} P)}{k_{-1} + k_2 + k_1 S + k_{-2} P} \tag{2}$$

The rate constant of Eq. (2) can be replaced by the constants of Michaelis-Menten equation of forward and reverse reaction (1).

$$v = \frac{E(V_{mf} K_{mr} S - V_{mf} K_{mf} P)}{K_{mr} K_{mf} + K_{mr} S + K_{mf} P} \tag{3}$$

where

$$V_{mf} = E_o k_2, \quad V_{mr} = E_o k_{-1}$$

$$K_{mf} = \frac{k_{-1} + k_2}{k_1}, \quad K_{mr} = \frac{k_{-1} + k_2}{k_{-2}}$$

The number of constants in Eq. 2 or 3 can be reduced by introducing a new concentration and equilibrium relation.

$$S_o + P_o = S + P \tag{4}$$

$$S_o + P_o = (K + 1) S_e \tag{5}$$

where

$$K = P_e/S_e = k_1 k_2/k_{-1} k_{-2} = V_{mf} K_{mr}/V_{mr} K_{mf} \qquad (6)$$

The reduced concentration of substrate, $\bar{S}$, will be defined such that it goes to zero as the reaction approaches equilibrium.

$$\bar{S} = S - S_e \qquad (7)$$

Substituting Eq. 4-7, we obtain the following time-course reaction rate equation in terms of reduced concentration, which is now formally identical to the ordinary Michaelis-Menten equation.

$$v = \frac{d\bar{S}}{dt} = \frac{EV'_m \bar{S}}{K'_m + \bar{S}} \qquad (8)$$

where

$$\left\{ V'_m = \left[\frac{K+1}{K} \right] \left[\frac{K_{mr}}{K_{mr} - K_{mf}} \right] V_{mf} \right. \qquad (9)$$

$$K'_m = \left[\frac{1}{K_{mr} - K_{mf}} \right] \left[(K_{mr} + K_{mf}K)S_e + K_{mr}K_{mf} \right] \qquad (10)$$

It will be noted that Eq. 8 is a generalized Michaelis-Menten equation for reversible reactions in which the irreversible reaction can be treated as a special case. (It can be easily shown that V'_m and K'_m in Eq. 8 become V_{mf} and K_{mf} as k_{-2} approached zero.) V'_m is still a quantity proportional to the enzyme concentration, and K'_m is a linear function of the initial substrate concentration. The symmetry of the reaction mechanism indicates that Eq. 8 is also applicable to reverse reactions with a position exchange of K_{mf}, K_{mr} and V_{mf}, V_{mr} in the definition of V'_m and K'_m.

Integration of Eq. 8 leads to the following relative conversion and time relation.

$$t = \left[\frac{S_o}{EV'_m} \right] X - \left[\frac{K'_m}{EV'_m} \right] \ln(1 - X) \qquad (11)$$

where

$$X = \frac{S_o - S}{S_o - S_e}$$

III.a. ENZYME IMMOBILIZATION

Acetone precipitated glucose isomerase from Strepto-myces sp. was a gift from a company who wishes its name to remain undisclosed and was used without further purification. The enzyme support material, ZrO_2 coated porous glass (96% silica, 40-80 mesh, 550 Å pore), was obtained from Corning Glass Works. The enzyme was covalently attached to the support by the method given by Messing and Weetall (2), which involves silanization of glass followed by glutaraldehyde linkage of this enzyme. The enzyme exhibited 4.7 units (μ-mole/min @ $60^{\circ}C$) in free form. The amount of enzyme offered for immobilization was 110 mg per g of glass. The assay of bulk enzyme solution before and after glass addition indicated 48 mg of the enzyme was attached per gram of glass. Comparison of free enzyme based activity of the bound enzyme and the activity of the enzyme-glass determined by direct assay showed that the enzyme retained 56% of the original activity after immobilization.

III.b. DETERMINATION OF REACTION RATE OR CONVERSION

A recirculated differential batch reactor system (shown in Fig. 1) was used for most of this study. Such a reactor system was probably first used by Beeck et.al. (3) and was recently used by Ford et.al. (4) for immobilized enzymes. Briefly, the system consists of a differential packed bed reactor (1.2 cm I.D.), a temperature equilibrating heat exchanger, a feed reservoir, and a peristaltic pump appropriately connected. The outlet stream of the packed-bed reactor was recirculated to the feed reservoir until the desired conversion was achieved. For close simulation of batch reaction, the single pass conversion in recycle stream was kept below 5% by adjusting bed height and flow rate. The reaction conversion or reaction rate was determined by intermittent sampling from the reactant reservoir and analysis of glucose content using Beckman Glucose Analyzer (Model EAR-2001). The immobilized enzyme was continuously reused until detectable deactivation had occurred. Since the samples were analyzed for glucose content

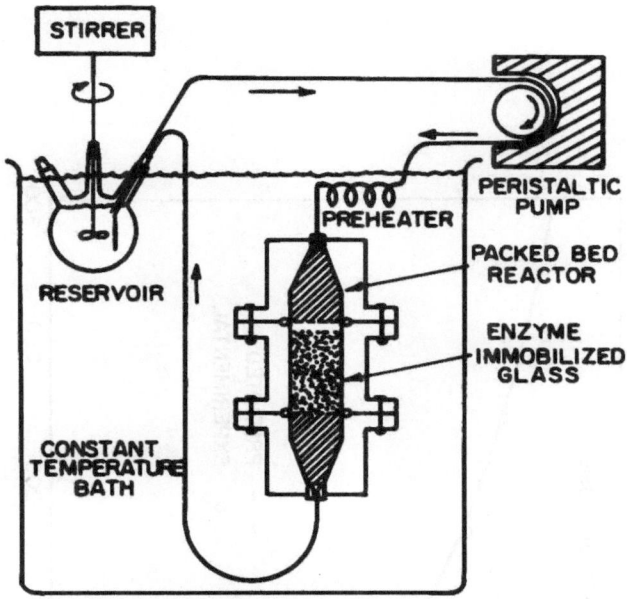

<u>Fig. 1</u>: Differential batch reactor with a packed bed column.

in both forward and reverse reactions, the accuracy in rate
determination was much higher for the reverse reaction
(fructose to glucose). Therefore, the experiments involving
mass transfer were conducted for the reverse reaction. The
substrate (glucose or fructose) exists in two isomers of
α and β form, and the enzymic reaction occurs to β form.
However, the mutarotion was substantially faster than the
enzymic reaction and thus did not influence the overall
reaction.

III.c. ACTIVATOR AND BUFFER

Magnesium and cobalt ions were known to have activating effects on the catalytic action of glucose isomerase
(5,6). The enhancing effects of these metal ions on the
activity of enzyme-glass prepared by the method identical
to that in this experiment, were studied by Fratzke (7).
The concentrations of cobalt and magnesium ions were kept
at the optimum levels of $10^{-4}M$ and $10^{-2}M$, respectively, in

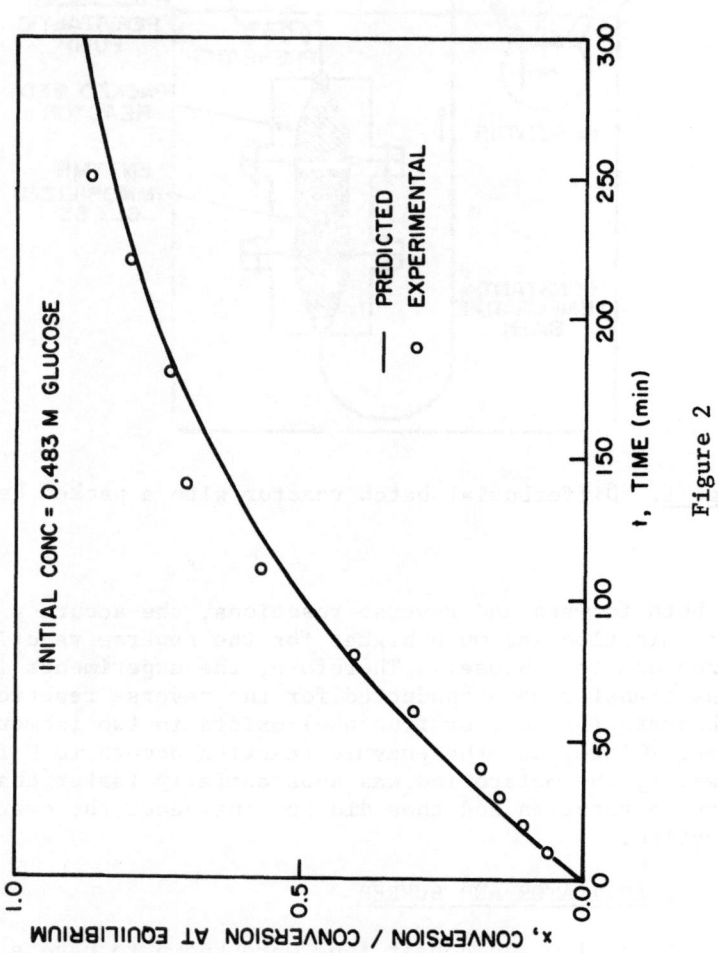

Figure 2

all the experiments. The solution pH was always preadjusted
to the optimum of 7.0 (7) with 0.02 M -glycerophosphate
buffer and hydrochloric acid.

IV. RESULTS AND DISCUSSION

IV.a. KINETICS

The constants in the time-course rate equation (Eqs.
9 and 10) are composed of the constants for the initial reaction rates of the forward and reverse reactions. The
first step in this study was to measure the initial forward
and reverse reaction rates. In the experiment, the conversions in forward and reverse reactions were traced up to
near the equilibrium point at five different substrate concentrations, ranging 0.018M to 1.80M. The reaction temperature was 50°C, and 50ml of the solution was recirculated
through a packed bed containing 3.5 ml (bulk volume) of
enzyme-glass at a superficial velocity of 1 cm/sec or above.
Under these experimental conditions the diffusional effects
on the overall reaction were considered to be negligible,
as is verified in a later part of this work. From the initial reaction rates taken from the same experiments the following constants were estimated through Lineweaver-Burk
plots (Fig. 2).

$$K_{mf} = 0.21 \text{ M}$$
$$K_{mr} = 0.40 \text{ M}$$
$$V_{mf} = 8.0 \text{ mg/(min.)(ml of enzyme-glass)}$$
$$V_{mr} = 15.7 \text{ mg/(min.) (ml of enzyme-glass)}$$

The Michaelis constants measured in this experiment were
considerably higher than the value of free enzyme reported
by Takasaki (6). The equilibrium constant observed in this
experiment was 1.03 at 50°C and appears to be in close agreement with the values of 0.92 at 40°C and 1.15 at 60°C also
reported by Takasaki. This value matched closely also with
the equilibrium constant of 1.02 calculated using Eq. 6 and
the initial reaction constants listed above.

The conversion data obtained here were in reasonable
agreement with the prediction of Eq. 11, the integrated con-
version-time relation, as shown in Figs. 2 and 3. Consistent underestimation of conversion at higher substrate

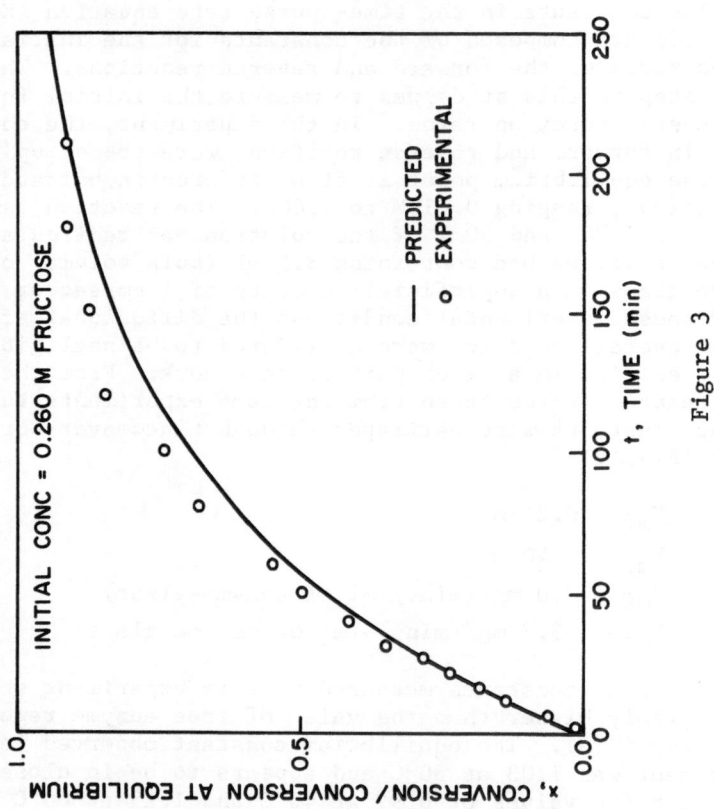

Figure 3

levels was noticed, but the discrepancy, with 10% maximum, was in the neighborhood of experimental errors, which also involve the kinetic parameter estimation. The Michaelis constant of this reaction is unusually higher or the reaction pattern is closer to first-order than ordinary enzymic reactions of low K_m. Therefore, unlike the case with common enzymic reaction wherein the substrate concentration is such higher than K_m, the backmixing in the reactor has stronger effect in conversion. For instance the plug flow reactor would give significantly higher conversion than CSTR for this immobilized enzyme reaction (12).

IV.b. IMMOBILIZED ENZYME STABILITY

The decay of immobilized enzyme activity under continuous operation was traced by intermittent assay of the output stream of the packed bed. From the semi-log plot of the enzyme activity vs. time the first order decay of enzyme activity was confirmed (Fig. 4). The half-life graphically determined from the plot was 240 days at 50°C. In extended operation, however, unexpected rapid deactivation was encountered after 30 days. It is believed that the reactor bed was heavily contaminated with microorganism.

IV.c. PORE DIFFUSION EFFECT

A legitimate way of verifying the pore diffusional effect is to compare the reaction rates for different sizes of catalyst supports. Such a test was made in this study using two different sizes of porous glass, one being 310 microns in mean diameter and the other being 20 microns. The smaller size particle was prepared by grinding the large enzyme-glass using a magnetic stirrer. During the course of grinding such precautions as low stirring rate, keeping the temperature low (below 10°C), and buffering the solution at pH 7 were necessary to prevent enzyme deactivation.

The effectiveness factor or the index of pore diffusion with the conventional definition (8) was simply the ratio of reaction rates observed with the two different sized particles since the diffusional effect in the smaller size particle was verified to be negligible, although the pore size was still much smaller than the ground enzyme glass. The reaction rate comparison was made for initial stage of reverse reaction for convenience and accuracy in the reaction rate measurements. The effectiveness factor

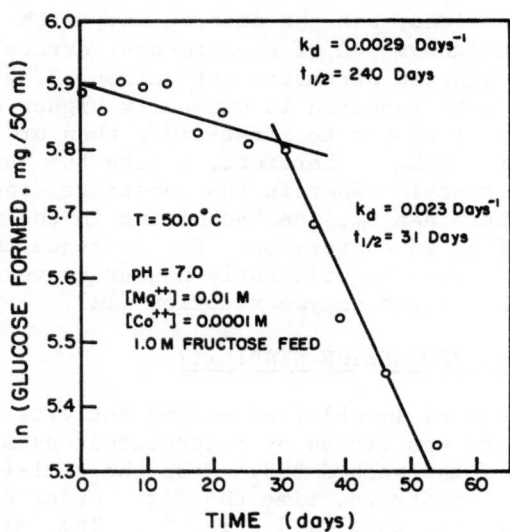

<u>Fig. 4</u>:   Semi-logarithmic plot of enzyme-glass activity
            versus time showing unexplained rapid deactiva-
            tion after 30 days of continuous operation--50°C.

determined at various temperatures over 50-75°C range and
at a substrate concentration of 0.5M indicated the pore
diffusion effect below 60°C was negligible (η at 60°C =
0.95).   These factors became increasingly important at
higher temperatures (η = 0.60 at 75°C).   In Fig. 5, the
measured effectiveness factors were plotted against the
Thiele modulus, $\phi_R = R\sqrt{V_m/D_eK_m}$, calculated at each tem-

perature using following parameters, either measured or
estimated.

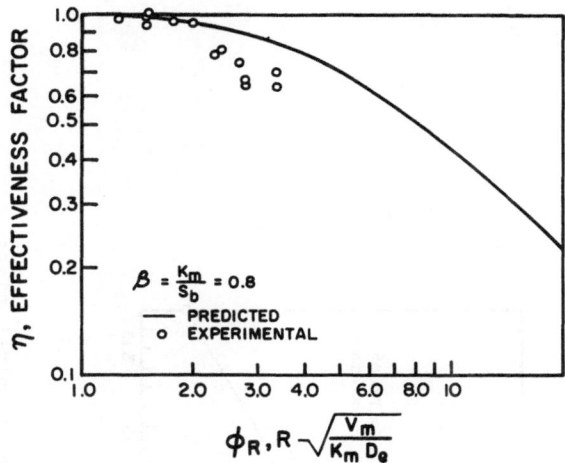

<u>Fig. 5</u>:   Comparison of experimental and theoretical effec-
tiveness factor.

$R = 0.016$ cm, $K_{mr} = 4 \times 10^{-4}$ mole/ml

$D_e = D\theta/\tau$ (9), $\theta = 0.79$, $\tau = 8.0$

$D = 6.9 \times 10^{-6}$ cm^2/sec at 50°C (corrected by Wilke-
Change equation for other temperatures)

$V_{mr} = 15.7$ mg/(min.) (ml of enzyme-glass) at 50°C
(Temperature correction was made by Arrhenius
equation with activation energy shown in the
following section.)

The solid line in Fig. 3 was transferred from the effectiveness factor chart for Michaelis-Menten kinetics prepared by Lee and Tsao (10). The measured effectiveness
factors were somewhat lower than the theoretical predictions
for the high temperature region, but the breaking point in
the η - ϕ_R relation closely agreed. The additional resis-

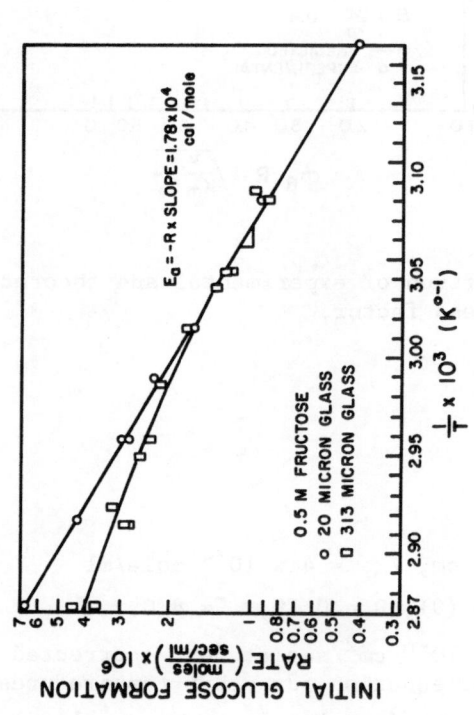

<u>Fig. 6</u>:   Arrhenius plot of reactions of two catalyst sizes.

tance at the external liquid film combined with some un-
certainties involved in the estimation of physical constants are cited as the reason.

The results shown here were based on the reverse
initial stage reaction. The forward reaction rate, as estimated by the kinetic constants obtained in the previous
section, is slightly lower than the reverse reaction, the
difference depending on the substrate concentration. Therefore, somewhat less pore diffusional effect is expected
for the forward reaction. For non first-order reactions
including Michaelis-Menten kinetics, the effectiveness factor depends not only on the Thiele modulus but also on the
substrate concentration. The situation of a decreasing
effectiveness factor at low substrate levels and its relation to integral or batch reactor behaviour was analytically
treated by Lee and Tsao (10).

From the reaction rate data obtained for effectiveness
factor measurements, Arrhenius plots were prepared for
ground and unground enzyme-glass (Fig. 6). In the plot,
the ground glass gave a straight line over the temperature
span $42^{\circ}C$ - $75^{\circ}C$, while the unground glass deviated from
the straight line for temperatures above $60^{\circ}C$, where the
pore diffusion resistance came into effect. The activation
energy of 17.8 Kcal estimated from the plot agreed quite
closely with the value reported by Takasaki et al (11).

IV.d. FILM DIFFUSION STUDY

In addition to the pore diffusion, an immobilized enzyme reaction is influenced by the substrate diffusion
through the external liquid layer, often termed liquid film.
The nature of the liquid film resistance is characterized
by its involvement with the fluid dynamic condition of the
system. Such a fact provides a means to qualitatively test
the importance of film resistance, as suggested by Levenspiel (13). In the packed-column case, the method of test-
ing is to observe the reaction rate at changing flow rates.

In this study, the same type of experiment was conducted using the recirculated differential, packed-bed reactor described earlier, and the decrease of reaction rate
at lower recirculation flow rates was observed (Fig. 7).
In all the tests conducted at four different temperatures,

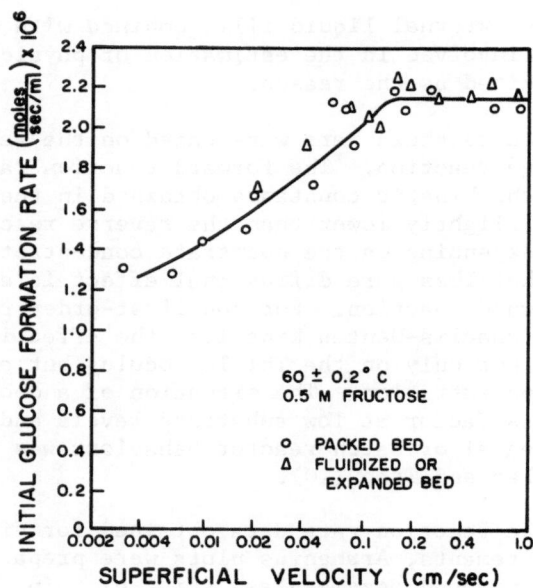

<u>Fig. 7</u>:   Comparison of film resistances between a packed-
bed and a fluidized-bed column.

rather sharp breaking points were noted. The flow rate at
this breaking point, referred to as critical flow rate,
u_c, increased with the temperature. This can easily be
explained by the fact that the reaction rate is more sensitive to temperature increase than the diffusion rate. The
critical flow rates beyond which the film diffusional effect becomes unimportant, observed in this experiment are
listed below in terms of superficial velocity (flow rate
divided by reactor cross-sectional area).

Temperature ($^{\circ}$C)	u_c, cm/sec
40	0.024
50	0.05
60	0.14
70	0.40

By reversing the flow direction in the reactor set up, a fluidized- or expanded-bed operation was performed for film resistance test. There was no observable difference between the two reactor beds in the reaction rates tested over flow rates ranging from 0.02 cm/sec to 1.0 cm/sec (superficial velocity). The results also indicate that there was no channeling in the packed bed, at least in this flow rate range and reactor size.

A theoretical treatment on film diffusion analogous to the pore diffusion case was made by Lee and Tsao (10) who derived the effectiveness factor for film diffusion, η_f, for Michaelis-Menten kinetics as follows.

$$\eta_f = \frac{\text{observed reaction rate with film resistance}}{\text{Reaction rate unhindered by film resistance}}$$

$$= \frac{(S_s/S_b)(1 + \beta)}{(S_s/S_b)\beta} \tag{12}$$

where

$$(S_s/S_b) = \frac{1}{2}[\{1 - \beta(\phi_f + 1)\} + \sqrt{\{1 - \beta(\phi_f + 1)\}^2 + \beta}] \tag{13}$$

$$\phi_f = \frac{LK_m}{k_L V_m}, \quad \beta = \frac{K_m}{S_b}$$

As Eqs. 12 and 13 indicate, the film diffusion effectiveness factor depends on two dimensionless parameters, ϕ_f, the film modulus, and β. The film effectiveness factors calculated with Eqs. 12 and 13 are shown in Fig. 8. This chart can be used for prediction of the film diffusional effect on the overall reaction knowing k_L, the mass transfer coefficient, which is included in the film modulus.

The following empirical correlations for k_L in a packed bed have been established in earlier work by McCune

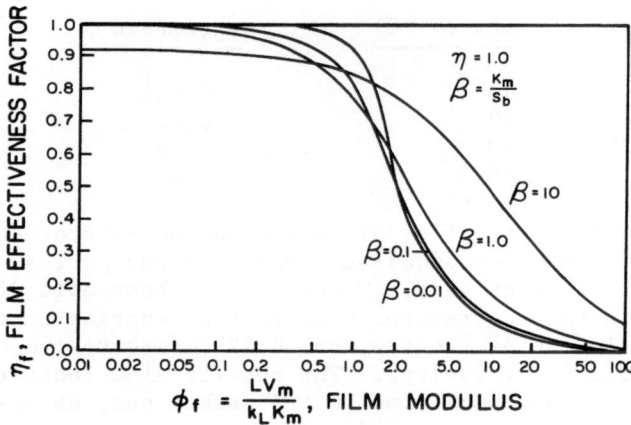

<u>Fig. 8</u>:   Film diffusion effectiveness factor chart.

and Wilhelm (14) and Wilson and Geankopolis (15).

$$\epsilon J = C (Re_p)^{-P} \tag{14}$$

where

$$J = (\frac{k_L}{u}) (\frac{\mu}{\rho D})^{2/3}, \ Re_p = \frac{d_p u \rho}{\mu}$$

	Re_p	C	P
Wilson and Geankopolis (1970)	0.0016 - 55	1.09	2/3
	55 - 1500	0.25	0.31
McCune and Wilhelm (1949)	8 - 20	1.625	0.507
	120 - 1300	0.687	0.327

The mass transfer coefficient, k_L, was estimated by employ-
ing the empirical correlation of Eq. 14.   The particle Reynolds number, Re_p, in this experiment was in the range 0.01
to 12, in which the Wilson and Geankopolis equation is applicable.   Then the film effectiveness factors, η_f, evaluate
from Eq. 12, where compared to the experimental studies of
film diffusion.   This comparison indicated that the pre-

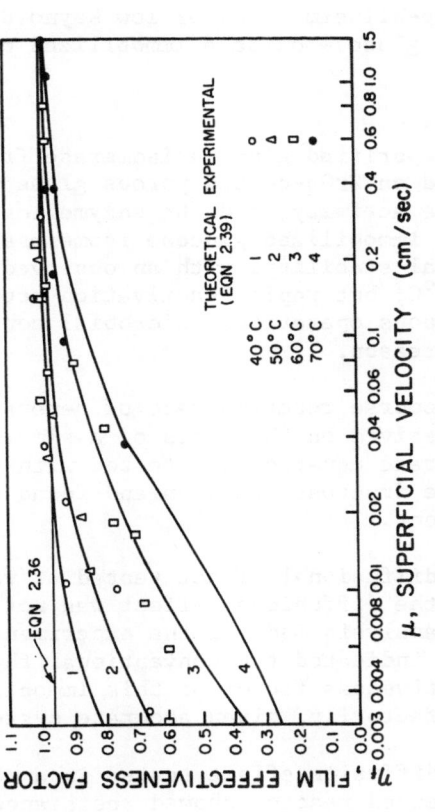

Fig. 9: Comparison of experimental and theoretical values of f.

dicted mass transfer coefficients were higher than the experimental values, by a factor of 2 to 4. In repeated comparison with other correlations listed above, the McCune-
Wilhelm model of higher Re_p (120~1300) was found to give
the closest agreement. The η_f predicted by the McCune-
Wilhelm model and the experimental η_f are shown in Fig. 9.
This result was not in accordance with the report by Rivitto and Kittrel (16), who found a satisfactory agreement
with the McCune-Wilhelm model of low Reynolds number in
their study of glucose oxidase immobilized on porous glass.

V. CONCLUSION

Partially purified glucose isomerase from streptomyces
sp. immobilized on ZrO_2-coated porous glass retained 56% of
the free enzyme activity, and the enzyme loading was 48 mg/g
of glass. The immobilized glucose isomerase exhibited remarkable thermal stability, with an observed half-life of
220 days at $50^\circ C$, but rapid deactivation occurred after 30
days of continuous operation. Microbial contamination was
cited for the reason.

The time-course reaction rate of reversible enzymic
reaction was derived on the basis of a simple reaction mechanism. The rate equation was tested with the conversion
data of glucose fructose reaction and found to be in reasonable agreement.

The pore diffusional effect tested at various temperatures showed the diffusional effect was not important below $60^\circ C$. An analysis made on the experimental results of
pore diffusion indicated the conventional theory would predict the effectiveness factor of this immobilized enzyme
reaction satisfactorily, given accurate system variables.

The film diffusion effect tested in a recirculated
differential packed reactor showed resistance of substrate
to diffusion through the external liquid film at extremely
low flow rates. The critical flow rates were determined at
several temperatures. No observable difference was found
in the film diffusional effect between packed-bed and fluidized-bed reactors. The channeling effect in a small packed-
column reactor was found to be negligible. The experimental
results on film diffusion studies were analyzed using the
empirical correlations of earlier workers and introducing

the film diffusion effectiveness factor. The Chilton-Colburn type correlation proposed by McCune-Wilhelm agreed more closely than others.

VI. ACKNOWLEDGMENT

This work was supported by the National Science Foundation through Grant GI-34933 and by the Engineering Research Institute at Iowa State University.

VII. REFERENCES

1. Walter, C., _Steady State Application in Enzyme Kinetics_, Academic Press, New York (1964).

2. Messing, R.A. and H.H. Weetall, US Patent #3515538 (1970).

3. Beeck, O., A.F. Smith and A. Wheeler, Proc. Royal Soc., A. _177_:62 (1940).

4. Ford, J.R., A.H. Lambert, W. Cohen and R.P. Chambers, _Enzyme Engineering_, Interscience Publishers, New York (1972).

5. Yamanaka, K., Biochem. Biophys. Acta. _151_:670 (1968).

6. Takasaki, Y., Agr. Biol. Chem. _31_:309 (1967).

7. Firatzke, A., M.S. Thesis, Chemical Engineering, Iowa State University (1974).

8. Smith, J.M, _Chemical Engineering Kinetics_, McGraw-Hill, New York (1970).

9. Satterfield, C.N., _Mass Transfer in Heterogeneous Catalysts_, MIT Press, p. 33 (1970).

10. Lee, Y.Y. and G.T. Tsao, J. of Food Science (in press).

11. Takasaki, Y., Y. Kosugi and A. Kaubayashi, Agr. Biol. Chem., _33_, 1527 (1969).

12. Lilly, M.O. and P. Dunnill, _Enzyme Engineering_, Edited by L.B. Wingard, Interscience Publishers, New York, p. 221 (1972).

13. Levenspiel, O., <u>Chemical Reaction Engineering</u>, John Wiley, New York (1973).

14. McCune, L.K. and R.H. Wilhelm, I & EC <u>41</u>:1124 (1949).

15. Wilson, E.J. and C.J. Geankopolis, I & EC, Fund. <u>5(1)</u>: 9 (1970).

16. Rivitto, B.J. and J.R. Kittrel, Biotech. Bioeng. <u>15</u>: 143 (1973).

<u>NOMENCLATURE</u>:

D = diffusivity
D_e = effective diffusivity
d_p = particle diameter
ϵ = bed voidage
E = enzyme or volume fraction of enzyme in the bed
E_o = enzyme concentration

$$J = (\frac{k_L}{u})(\frac{\mu}{\rho D})^{2/3}, \text{ dimensionless}$$

K = equilibrium constant
k_i = reaction rate constant
k_L = mass transfer coefficient at the liquid film
K_m' = Michaelis constant for reversible reaction, defined by Eq.
K_{mr} = Michaelis constant for forward reaction
K_{mf} = Michaelis constant for reverse reaction
L = characteristic length of enzyme support, exterior surface area/volume of enzyme support
P = product or product concentration
P_e = product concentration at equilibrium
Re_p = particle Reynolds number $\frac{d_p u \rho}{\mu}$, dimensionless
R = radius of enzyme support
S = substrate or substrate concentration
$\bar{S} = S - S_e$
S_e = substrate concentration at equilibrium
S_o = initial substrate concentration
$\bar{S}_o = S_o - S_e$

S_s = substrate concentration at outer surface of enzyme
        support
S_b = bulk substrate concentration
t = time
u = superficial velocity (flow rate/reactor cross-sectional
        area)
u_c = critical superficial velocity
v = reaction rate
v_{mf} = maximum reaction rate for forward reaction based on
        the volume of immobilized enzyme
V_{mr} = maximum reaction rate for reverse reaction based on
        the volume of immobilized enzyme
V_m' = maximum reaction rate for reversible reaction based
        on the volume of immobilized enzyme
X = relative concession, $(S_o - S)/(S_o-S_e)$
β = dimensionless Michaelis constant, K_m/C_b
η = pore diffusion effectiveness factor
η_f = film diffusion effectiveness factor
μ = viscosity
ρ = density
τ = tortuosity factor of enzyme support pores, dimensionless

ϕ = Thiele modulus, $R \sqrt{\dfrac{V_m}{D_e K_m}}$

ϕ_f = film modulus $\dfrac{L V_m}{k_L K_m}$

EFFECT OF MIXING AND MASS TRANSFER ON THE GLUCOSE ISOMERIZATION BY ENTRAPPED CELLS IN A VERTICAL PLATE TYPE REACTOR

Hisaharu Taguchi, Kennichi Suga, Toshiomi Yoshida
and Sadayuki Yuda
Dept. of Ferment. Technol., Faculty of Eng.
Osaka University, Yamada-kami, Suita-shi, Osaka
565, Japan

INTRODUCTION

In recent years, the importance and application of entrapped whole cells have been so frequently proposed, discussed, put into practice, and reviewed (1-3) as to need no
further reemphasis here.

However, in general when a packed bed reactor with gel
particles is used a decrease in the interfacial area occurs
during its operation due to changes in the physical characteristics of these particles, such as changes in volume and
form. As a consequence, a decrease in the overall yield
and an increase in pressure drop are observed. In addition,
in one particular case gas formation was observed, and un-
favorable hydrodynamic phenomena were caused by the rising
counter-current of gas bubbles.

To eliminate these difficulties a vertical gel plate
type reactor was used in this work and applied to the entrapping of Streptomyces sp. cells containing active glucose isomerase. The use of this type of reactor also sim-
plifies the analysis of the basic characteristics of gel
entrapped cells systems.

The effect of mixing and mass transfer on the degree
of conversion were studied both experimentally and theoret-
ically, and simulation studies were also performed.

151

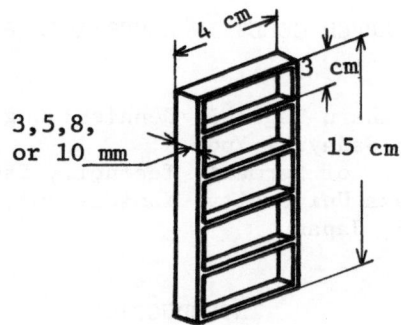

Fig. 1:   Diagram of plate frame.

EXPERIMENTAL

The gel plates were prepared by the following procedure (4): to 4 ml of preheated Streptomyces sp. cell suspension (about 0.5 weight %), 750 mg of acrylamide monomer
and 40 mg of N,N'-methylenebisacrylamide were added; subsequently 0.5 ml of 5% 3-dimethylamino propionitrile and 0.5
ml of 2.5% potassium persulphate solution were added. The
reaction mixture was poured into the five cells of the plate
frames shown in Fig. 1. The polymerization was completed by
further incubation at 37°C for 30 min. Figure 2 shows a
schematic diagram of the continuous plate type reactor system which consists of a jacketed polypropylene reactor (20

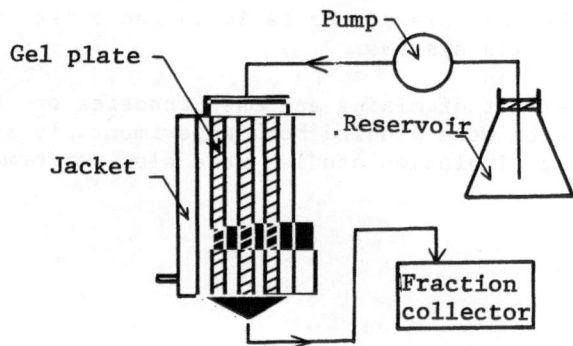

Fig. 2:   Schematic diagram of continuous plate type reactor.

cm high, 5 cm wide and 8 cm long), a preheated reservoir
for glucose solution, a peristaltic pump and a fraction
collector.

The number of plates within the reactor depends on
their thickness. 7,4 and 2 plates were used in cases where
the thickness was 3, 5 and 10 mm respectively. The height
of 15 cm and the width of 4 cm were constant.

The feed solution was prepared such that the concentration of its components were: 2.2 M D-glucose, 0.05 M MgSO$_4$,
0.024 M CoCl$_2$ and 0.1 M phosphate buffer (pH 7.0). D-fructose produced was determined by the cystine-carbazole method
and confirmed by measuring the remaining glucose using a
gluco-stat.

One unit of the enzyme activity was defined as the
amount of enzyme that produces one mg of D-fructose during
one hour. The maximum rates of consumption and formation
by preheated cells (V_A', V_B') and the Michaelis-Menten constants (K_A, K_B) were determined by using the values of the
initial rates of formation and consumption obtained in batch
reaction.

The flow reactor runs were carried out at temperatures
between 60°C and 75°C, with the flow rate of glucose solution varying from 30 to 90 cm^3/hr.

The mixing characteristics, particularly the values of
the dispersion coefficient (E), were determined using the
dispersion model (5).

The effective diffusion coefficient for glucose was
determined from the results obtained in an H-type cylindrical tube (0.7 cm diameter, 1.5 cm length). The cross-tube
was filled with inactivated cells entrapped in acrylamide
gel. Glucose solution (1 M) and water were flowed in each
vertical tube at the same rate, and glucose concentration in
the water stream was measured at steady state. The effective diffusion coefficient was calculated by the method suggested by Weitze (6).

THEORETICAL

Consider a reversible reaction involving the intercon-
version of glucose (A) and fructose (B) catalyzed by glucose

isomerase (7) (E), in which it is assumed that two inter-
mediate complexes, EA and EB, are formed:

$$A + E \underset{k_{-1}}{\overset{k_1}{\rightleftharpoons}} EA \underset{k_{-2}}{\overset{k_2}{\rightleftharpoons}} EB \underset{k_{-3}}{\overset{k_3}{\rightleftharpoons}} E + B \tag{1}$$

where k_1, k_2 and k_3 are the rate constants for the forward
processes and k_{-1}, k_{-2} and k_{-3} are the rate constants for
the reverse processes. According to Peller and Alberty's
work (8), at the steady state the rate of disappearance of
A, $-dC_A/dt$, or the rate of formation of B, dC_B/dt are given
as follows:

$$r = -\frac{dC_A}{dt} = \frac{dC_B}{dt} = \frac{\dfrac{V_A}{K_A} C_A - \dfrac{V_B}{K_B} C_B}{1 + \dfrac{C_A}{K_A} + \dfrac{C_B}{K_B}} \tag{2}$$

where C represents sugar concentration, and subscript A
denotes glucose and B fructose. V is the maximum reaction
rate in the reactor, and the correlation between V and V'
is given as follows:

$$V_A = V_A' \frac{\text{entrapped cell mass (g)}}{\text{volume of gel plate (ml)}} \tag{3}$$

Consider a plate reactor (plate height, L and width,
a) through which fluid flows with an average velocity, u,
and in which substrate and product mix longitudinally with
a dispersion coefficient, E.

For the analysis of this system the working hypotheses
are: 1.) Enzymatic activity is constant and homogeneous
along the plate. 2.) Diffusion of substrate and product
within the plate can be represented by Fick's first law;
the overall effective diffusion coefficient is constant.
3.) Longitudinal transfer of the substrate and product inside the plate is negligible. 4.) The concentration is substrate and product in the bulk flow is uniform over a cross
section, and mass transfer resistance between the external
solution and the plate can be neglected.

The substrate and product are transported by diffusion to and from differential section within the gel plate where the consumption or formation reaction occurs. At steady state, material balance over an element of volume (dx·dl·a) at a depth 1 to 1+dl yields:

$$D_A \frac{d^2 C_A}{dx^2} = r \tag{4}$$

$$D_B \frac{d^2 C_B}{dx^2} = - r \tag{5}$$

where D_A and D_B are the effective diffusion coefficient of substrate and product in the gel plate and x is the horizontal distance from the center of the plate.

The boundary conditions are given by the following Eqs:

$$\text{at } x=0 \quad \frac{dC_A}{dx} = 0, \quad \frac{dC_B}{dx} = 0 \tag{6}$$

$$\text{at } x=X \quad C_A = C_{AS}, \quad C_B = C_{BS} \tag{7}$$

where C_{AS} and C_{BS} are the sugar concentrations in the bulk solution. Eqs. (4) and (5) lead to

$$D_A \frac{d^2 C_A}{dx^2} = \frac{\dfrac{V_A}{K_A} C_A - \dfrac{V_B}{K_B} C_B}{1 + \dfrac{C_A}{K_A} + \dfrac{C_B}{K_B}} + \tag{8}$$

$$D_B \frac{d^2 C_B}{dx^2} = - \frac{\dfrac{V_A}{K_A} C_A - \dfrac{V_B}{K_B} C_B}{1 + \dfrac{C_A}{K_A} + \dfrac{C_B}{K_B}} \tag{9}$$

From Eqs. (8) and (9) the following correlation can be introduced.

$$D_A \frac{d^2 C_A}{dx^2} = - D_B \frac{d^2 C_B}{dx^2} \tag{10}$$

Integrating Eq. (10) from x=0 to x=x, the boundary conditions given by Eq. (6), yields,

$$D_A \frac{dC_A}{dx} = - D_B \frac{dC_B}{dx} \tag{11}$$

Reintegrating Eq. (11) from x=x to x=X, the boundary conditions given by Eq. (7), leads to

$$D_A (C_{AS} - C_A) = - D_B (C_{BS} - C_B) \tag{12}$$

Material balance over a differential cross section of the bulk flow yields:

$$E \frac{d^2 C_{AS}}{d\ell^2} - u \frac{dC_{AS}}{d\ell} = \frac{2nD_A a}{S} \frac{dC_A}{dx} \bigg|_{x=X} \tag{13}$$

$$E \frac{d^2 C_{BS}}{d\ell^2} - u \frac{C_{BS}}{d\ell} = \frac{2nD_B a}{S} \frac{dC_B}{dx} \bigg|_{x=X} \tag{14}$$

where   is the number of gel plates, and S is the area of the cross section of bulk flow. As given by Danckwert's expression (9), the boundary conditions are:

$$\text{at } \ell=0, \quad \left.\begin{array}{l} C_{AS} - (E/u)\,(dC_{AS}/d\ell) = C_{A0} \\[2mm] C_{BS} - (E/u)\,(dC_{BS}/d\ell) = C_{B0} \end{array}\right\} \tag{15}$$

$$\text{at } \ell=L, \quad \frac{dC_{AS}}{d\ell} = 0 \,, \quad \frac{dC_{BS}}{d\ell} = 0 \tag{16}$$

where C_{AO} and C_{BO} are the concentrations of the sugars in the feed solution. Summation of Eqs. (13) and (14), and integration from 1=0 to 1=L with Eqs. (15) and (16) as the boundary conditions, gives Eq. (17)

$$- C_{AS}\big|_{\ell=L} + C_{AO} - C_{BS}\big|_{\ell=L} + C_{BO} = 0 \tag{17}$$

Eqs. (8) and (9) can be expressed in dimensionless form as follows:

$$\frac{d^2\omega}{d\eta^2} = \hat{\delta}\,\frac{\omega - \alpha\nu}{1+\beta\omega+\gamma\nu} \tag{18}$$

$$\frac{d^2\nu}{d\eta^2} = -\delta'\,\frac{\omega - \alpha\nu}{1+\beta\omega+\gamma\nu} \tag{19}$$

and the boundary conditions are

$$\text{at } \eta = 0, \quad \frac{d\omega}{d\eta} = 0, \ \frac{d\nu}{d\eta} = 0 \tag{20}$$

$$\text{at } \eta = 1, \quad \omega = \omega_s, \quad \nu = \nu_s \tag{21}$$

Eq. (12) in the same manner becomes:

$$\nu = \nu_s + \zeta\,(\omega_s - \omega) \tag{22}$$

where

$$\omega = \frac{C_A}{C_{AO}}, \quad \omega_s = \frac{C_{AS}}{C_{AO}}, \quad \nu = \frac{C_B}{C_{AO}}, \quad \nu_s = \frac{C_{BS}}{C_{AO}}$$

$$\alpha = \frac{V_B}{K_B}\frac{K_A}{V_A}, \quad \beta = \frac{C_{AO}}{K_A}, \quad \gamma = \frac{C_{AO}}{K_B}, \quad \delta = \frac{\chi^2}{D_A}\frac{V_A}{K_A},$$

$$\eta = \frac{x}{X}, \quad \zeta = \frac{D_A}{D_B} \text{ and } \delta' = \frac{\chi^2}{D_B}\frac{V_B}{K_B}$$

Furthermore Eqs. (13) and (14) can be rewritten in the dimensionless form as follows:

$$\frac{d^2\omega_s}{d\xi^2} - \varepsilon\frac{d\omega_s}{d\xi} = \sigma\frac{d\omega}{d\eta}\bigg|_{\eta=1} \tag{23}$$

$$\frac{d^2\nu_s}{d\xi^2} - \varepsilon\frac{d\nu_s}{d\xi} = \sigma'\frac{d\nu}{d\eta}\bigg|_{\eta=1} \tag{24}$$

The boundary conditions are:

$$\text{at } \xi=0, \qquad \omega_s - \frac{1}{\varepsilon}\frac{d\omega_s}{d\xi} = 1 \tag{25}$$

$$\nu_s - \frac{1}{\varepsilon}\frac{d\nu_s}{d\xi} = \kappa$$

$$\text{at } \xi=1, \qquad \frac{d\omega_s}{d\xi} = 0, \quad \frac{d\nu_s}{d\xi} = 0 \tag{26}$$

where

$$\sigma = \frac{2\,naD_A L^2}{SEX} \quad, \quad \sigma' = \frac{2\,naD_B L^2}{SEX} \quad, \quad \varepsilon = \frac{uL}{E} \quad,$$

$$\xi = \frac{\ell}{L} \quad \text{and} \quad \kappa = \frac{C_{BO}}{C_{AO}}$$

The exit concentration of fructose could be calculated numerically from Eqs. (18), (22), (23) and (24) using the experimental values of the mixing characteristics of the flow system, reaction rate and effective diffusion coefficient of glucose and fructose in the gel plate.

As is well known, the concept of the effectiveness factor, which has been exhaustively discussed by several authors (10-12), is of great benefit in the case of first order reactions, when the modulus becomes independent of the surface reactant concentration, or in cases where this surface concentration is constant. However in cases of other systems, such as the complex enzymatic reaction of glucose isomerization in this reactor with non-ideal flow characteristics, over-all efficiency of the system can not be esti-

mated directly from the effectiveness factor. In view of
this fact, in the present paper, to study the effect of
mixing in the bulk flow and diffusion in the plate on the
degree of conversion, the efficiency of the reactor (E.R.)
defined as follows was applied.

$$\text{E.R.} = \frac{\text{Actual conversion in this reactor}}{\text{Maximum conversion in an ideal reactor}}$$

Maximum conversion might be obtained in the plug flow reactor where diffusion resistance is negligible. Actual conversion was obtained by solving the above equations numerically. The applicability of this concept of efficiency of
reactor need not be limited to this case, but can be extended to other types of reactors involving chemical or biochemical reactions.

If there were no effect of diffusion on the reaction
and the mixing characteristic corresponded to that in ideal
flow, i.e., plug flow or backmix, the conversion could be
obtained analytically.

1.) In the case of plug flow
    Material balance over a differential cross section of
the bulk flow yields:

$$-uS \ \frac{dC_{AS}}{d\ell} = \frac{\dfrac{V_A}{K_A} C_{AS} - \dfrac{V_B}{K_B} C_{BS}}{1 + \dfrac{C_{AS}}{K_A} + \dfrac{C_{BS}}{K_B}} \ 2aXn \tag{27}$$

$$-uS \ \frac{dC_{BS}}{d\ell} = \frac{-\dfrac{V_A}{K_A} C_{AS} + \dfrac{V_B}{K_B} C_{BS}}{1 + \dfrac{C_{AS}}{K_A} + \dfrac{C_{BS}}{K_B}} \ 2aXn \tag{28}$$

From Eqs. (27) and (28) we obtain

$$\frac{dC_{AS}}{d\ell} + \frac{dC_{BS}}{d\ell} = 0 \tag{29}$$

Integration from 1=0 to 1=1 gives

$$C_{AO} - C_{AS} + C_{BO} - C_{BS} = 0 \tag{30}$$

Eq. (27) can be rewritten in dimensionless form as follows:

$$- \frac{d\omega_s}{d\xi} = \theta \frac{\omega_s - \alpha(1-\omega_s)}{1+\beta\omega_s+\gamma(1-\omega_s)} \tag{31}$$

where

$$\theta = \frac{LV_A}{uK_A} \cdot \frac{2aXn}{S}$$

Hence, integration of Eq. (31) from $\xi = 0$ to $\xi = \xi$, yields

$$\lambda ln\{(1 + \alpha)\omega_s - \alpha\} + (\beta - \gamma)(\omega_s - 1) = - (1 + \alpha)\ \theta\xi \tag{32}$$

where

$$\lambda = (1+\gamma) + \frac{\alpha(\beta-\gamma)}{1 + \alpha}$$

2.) <u>In the case of backmix</u>
    Material balance in the steady-state backmix flow reactor gives

$$uS\ C_{AO} - uS\ C_{AS} = \frac{\frac{V_A}{K_A} C_{AS} - \frac{V_B}{K_B} C_{BS}}{1 + \frac{C_{AS}}{K_A} + \frac{C_{BS}}{K_B}}\ 2aXLn \tag{33}$$

$$uS\ C_{BO} - uS\ C_{BS} = \frac{- \frac{V_A}{K_A} C_{AS} + \frac{V_B}{K_B} C_{BS}}{1 + \frac{C_{AS}}{K_A} + \frac{C_{BS}}{K_B}}\ 2aXLn \tag{34}$$

Eq. (33) can be expressed in dimensionless form as

$$1 - \omega_s = \theta \frac{\omega_s - \alpha(1 - \omega_s)}{1 + \beta\omega_s + \gamma(1-\omega_s)} \tag{35}$$

$$\omega_s = \frac{\beta-2\gamma-1-(1+\alpha)\theta + \sqrt{\{1+(1+\alpha)\theta+2\gamma-\beta\}^2 + (\beta-\gamma)(2\theta+1+\gamma)}}{2(\beta - \gamma)} \tag{36}$$

TABLE 1

EFFECTIVE DIFFUSION COEFFICIENT

Temperature (^{o}C)	D_A (cm^2/hr)
50	1.29×10^{-2}
60	1.96×10^{-2}
70	2.29×10^{-2}

RESULTS AND DISCUSSION

1.) Effective Diffusion Coefficient and Dispersion Coefficient

Measurement at various temperatures of the effective diffusion coefficient of the sugars through the gel in which inactivated cells were entrapped was carried out and the results are shown in Table 1. No marked differences between the effective coefficient of fructose and glucose were observed.

The dispersion coefficients in the reactor (bulk flow) at flow rates of 30, 60 and 90 ml/hr were 4.99, 7.21 and 8.22 cm^2/hr respectively.

2.) Kinetic Constant of Glucose Isomerization by Preheated Cells

The maximum rates of consumption and formation, and the Michaelis-Menten constants of preheated cells obtained in batch reactions at various temperatures are listed in Table 2. The values of the maximum rates of consumption and formation increase with the increase in temperature, as expected. However, the increase of the values of K_A with temperature and the apparent invariance of the K_B values with temperature changes are phenomena that, so far, we can not explain.

3.) Simulation and Computation

Figure 3 illustrates typical profiles of the glucose concentration in the bulk flow in the longitudinal direction

TABLE 2

KINETIC CONSTANTS OF GLUCOSE ISOMERIZATION BY PREHEATED CELLS AT VARIOUS TEMPERATURES

Kinetic constant	Temperature				
	60°C	65°C	70°C	75°C	80°C
V_A' (mol/hr·g cell)	8.8×10^{-3}	1.47×10^{-2}	3.42×10^{-2}	4.87×10^{-2}	7.20×10^{-2}
V_B' (mol/hr·g cell)	1.98×10^{-2}	2.92×10^{-2}	3.38×10^{-2}	3.93×10^{-2}	6.60×10^{-2}
K_A (mol/hr)	0.18×10^{-3}	0.21×10^{-3}	0.35×10^{-3}	0.39×10^{-3}	0.54×10^{-3}
K_B (mol/hr)	0.32×10^{-3}	0.37×10^{-3}	0.35×10^{-3}	0.32×10^{-3}	0.42×10^{-3}
K^* (-)	0.79	0.88	1.01	1.02	0.85

* $K = (V_A/K_A)/(V_B/K_B)$

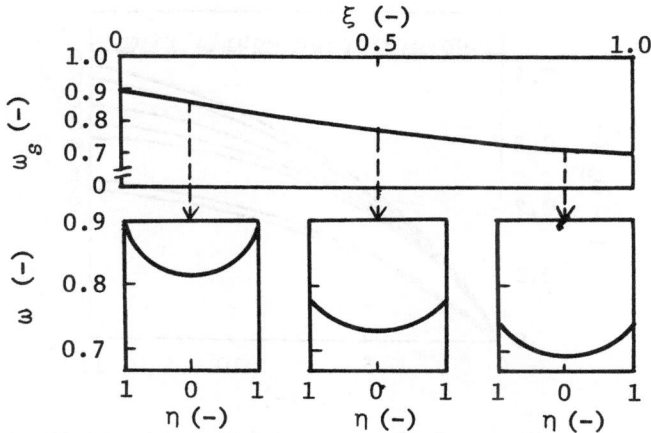

<u>Fig. 3</u>:   Profiles of glucose concentration in the reactor
          and in the gel plate. The upper figure shows the
          longitudinal glucose distribution in bulk flow;
          the lower figures show glucose distribution in the
          gel plate at different reduced distances from the
          inlet. Gel thickness: 10 mm. Operating tempera-
          ture: 60°C. Flow rate: 60ml/hr.

as well as in the gel plates at different reduced distances
from the inlet, as calculated from the model. The dependency of the conversion ratio on the θ values is shown in
Fig. 4. The effect of the degree of backmixing in bulk
flow when the diffusion resistance in the gel plate is negligible can be visualized by comparing the lines a, b and
d. It is obvious that a higher conversion rate is attained
at a lower degree of backmixing. On the other hand, the effect of diffusion resistance in the gel plate at a certain
degree of backmixing (ϵ =3.76) can be visualized comparing
lines b, c and f. It is clear that a higher diffusion resistance gives a lower conversion.

     Further, comparing the results of the actual process
ϵ =3.76) with the ideal case (ϵ =0), i.e., line c with e
and line f with g, we can conclude that at higher values of
δ , corresponding to higher diffusion resistances, the effect of the degree of backmixing is comparatively reduced.

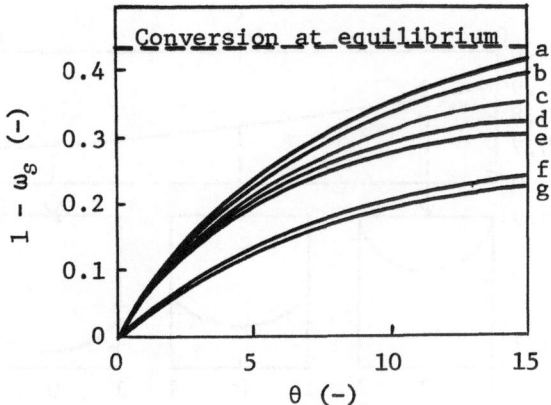

Fig. 4: Dependency of the conversion ratio on the value of θ under different mixing conditions and diffusion resistances.
The values of   and   for each line:

Line	a	b	c	d	e	f	g
	0	0	3.88	0	3.88	43.13	43.13
		3.76	3.76	0	0	3.76	0

For any desired conversion ratio, the corresponding θ value can be obtained from Fig. 4, enabling the determination of the respective residence time or the amount of pre-heated cells to be entrapped. Furthermore, knowing the θ value, the efficiency of the reactor, E.R., can be obtained from Fig. 5 and Fig. 6 for the reactions at $60^{\circ}C$ and $75^{\circ}C$ respectively.

The results mentioned previously demonstrate the marked effect of diffusion inside the plate on the efficiency of the reactor. In fact, for a 3 mm gel plate (δ =3.88), θ value is 13.5 (residence time 13.1 hr), the conversion ratio is 0.35, and the E.R. obtained from Fig. 5 is 0.84. For a 10 mm gel plate, the E.R. obtained in the same manner has a rather lower value of 0.51. Comparing Fig. 5 and Fig. 6, at the higher temperature, the retention time corresponding to the minimum E.R. can be observed more clearly.

The experimental values as well as the calculated values obtained from the proposed model are listed in Table 3.

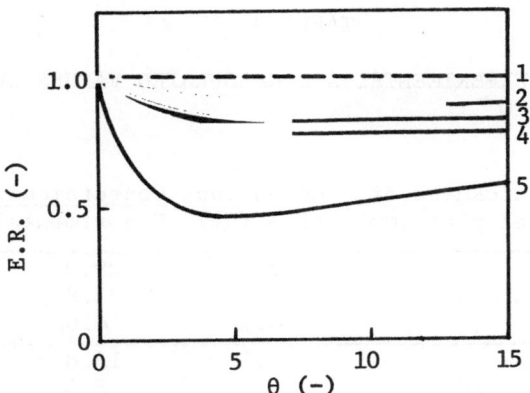

Fig. 5: Dependency of the reactor efficiency on the value
of θ under different mixing conditions and diffusion resistance. Operating temperature: 60°C.
The values of   and   for each line:

Line	1	2	3	4	5
	0	0	3.88	0	30.90
		3.76	3.76	0	3.76

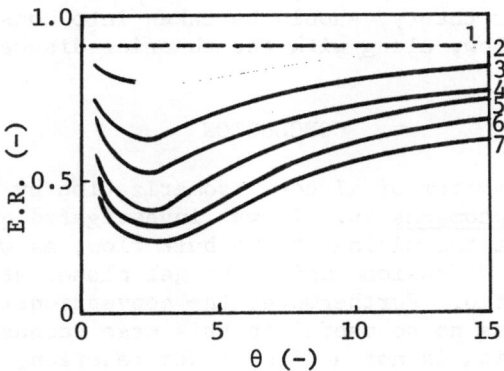

Fig. 6: Dependency of the reactor efficiency on the value of
θ under different mixing conditions and diffusion
resistance. Operating temperature: 75°C.

Line	1	2	3	4	5	6	7
	0	6.97	6.97	6.97	77.41	77.41	77.41
			2.0	0		2.0	0

TABLE 3

COMPARISON OF EXPERIMENTAL AND CALCULATED VALUES OF THE
CONVERSION RATIO

Temperature (°C)	Thickness of gel plate (mm)	Retention time (hr)	Conversion ratio (%)	
			Experimental	Calcula-ted
60	3	13.1	29.9	35.1
		6.8	23.8	28.6
		4.2	14.6	19.9
	10	6.3	13.8	18.0
		4.2	8.4	12.3
70	3	12.1	41.0	48.6
75	3	13.7	41.2	48.8

It is observed that the calculated values are slightly
higher than the experimental values. This difference might
be due to two main reasons: the scraping of cells from the
gel plate, and the resistance to mass transfer between the
bulk flow and the plate. Mention should be made here that,
to minimize this difference, other theories, such as the
boundary layer theory, should be taken into consideration
in the bulk flow, along with the theories discussed in the
present work.

CONCLUSION

In this system of glucose isomerization by entrapped
cells of Streptomyces sp., it was investigated quantitatively
to what extent the mixing of the bulk flow, as well as the
resistance to diffusion inside the gel plate, affected the
conversion ratio. Furthermore, the conventional effectiveness factor was no so useful in this case because this enzymatic reaction is not a first order reaction, and sugar
concentration in the surface of gel plate is not uniform in
the longitudinal direction. So a new concept of efficiency
of reactor, E.R. was introduced to enable the determination
of the characteristics of this reactor. Comparison between
the experimental and calculated values indicates that the
proposed model accounts quite well for the actual performance
of this system.

REFERENCES

1. Silman, I.H., Katchalski, E., Ann. Rev. Biochem. <u>35</u>,
    87 (1966).

2. Hattori, T., Furusaka, C., J. Biochem. <u>50</u>, 312 (1961).

3. Johnson, D.E., Ciegler, A., Arch. Biochem. Biophys.
    <u>130</u>, 384 (1969).

4. Wieland, T., Determen, H., Bünnig, K., Z. Naturforsch.
    <u>216</u>, 1003 (1966).

5. Levenspiel, O., "Chem. Reaction Eng.", p. 261, John
    Wiley and Sons, Inc., New York (1962).

6. Weize, P.B., Z. Physik. Chem. Folge. <u>11</u>, 1 (1957).

7. Takasaki, Y., Report Ferm. Res. Ins. <u>36</u>, 45 (1969).

8. Peller, L., Alberty, R.A., J. Am. Chem. Soc. <u>20</u>, 5907
    (1959).

9. Danckwerts, P.V., Chem. Eng. Sci. <u>2</u>, 1 (1953).

10. Bischoff, K.B., A.I.Ch.E. J. <u>11</u>, 351 (1965).

11. Hougen, O.A., Ind. Eng. Chem., <u>53</u>, 509 (1961).

12. Moo-Young, M., Kobayashi, T., Canadian J. Chem. Eng.
    <u>50</u>, 162 (1972).

IMMOBILIZATION OF β-GALACTOSIDASE BY POLYACRYLAMIDE GEL

Takeshi Kobayashi, Kunio Ohmiya and Shoichi
Shimizu
Department of Food Science and Technology
Faculty of Agriculture, Nagoya University
Nagoya 464, Japan

Milk intolerance resulting from a deficiency of intestinal lactase (β-galactosidase) is now a well-defined clinical syndrome in both infants and adults (1) and prevails much higher among Orientals and Negroes than Caucasians (2). A number of Japanese are also reported to be milk-intolerant (3). Furthermore, intestinal malabsorption of lactose in infancy and childhool makes the treatment of refractory diarrhea difficult (4). To avoid these troubles, the efficient and inexpensive processing of milk to reduce lactose content is desirable.

Recently, such processing has been reported by applying immobilized β-galactosidase (5,6,7,8,9,10). In the present investigation, immobilization of β-galactosidase by polyacrylamide gel was carried out in the presence of protective agents and enzyme beads with high activity were obtained. Some characteristics of the enzyme immobilized by this procedure were compared with those in free state.

PREPARATION AND CHARACTERISTICS OF IMMOBILIZED ENZYME

Three kinds of β-galactosidase were used in the present investigation. β-Galactosidase from <u>Aspergillus oryzae</u> was a gift of Tokyo Tanabe Co., Ltd. Other kinds of β-galactosidase were prepared from <u>Escherichia coli</u> K12 and <u>Kluyveromyces</u> (<u>Saccharomyces</u>) <u>lactis</u> as follows. Microorganisms cultured in their suitable media were harvested by centrifugation and disrupted with powder of alminium oxide. Crude enzyme was extracted with M/15 phosphate buffer (pH 6.5), partially purified by salting-out with ammonium sul-

169

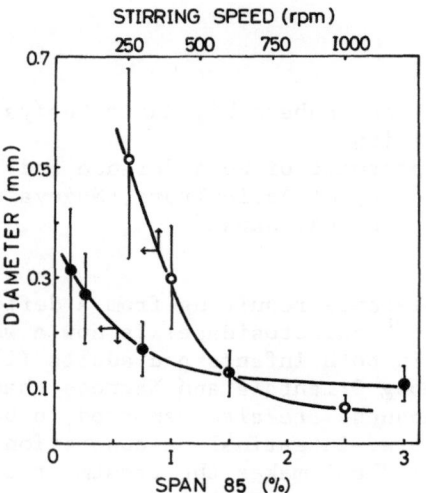

<u>Fig. 1</u>:   Relationship between diameter of beads and concen-
        tration of Span 85 (stirring speed : 600 rpm) or
        stirring speed (concentration of Span 85 : 1.5%).

fate (20 - 80% saturation), dialyzed and lyophilized.

    <u>Preparation of polyacrylamide gel beads</u>. Preparation
of polyacrylamide gel beads containing β-galactosidase was
essentially the same as that reported by Dahlqvist et al
(5) except that N,N,N',N'-tetramethyl-ethylenediamine (TEMED)
was modified to add to an organic solvent. β-galactosidase
and its protective agents were dissolved in M/15 phosphate
buffer (pH 6.5), 3.5 ml. This enzyme-protective agent mixture was added to the same volume of aqueous solution containing 25% acrylamide and N,N'-methylene-bis-acrylamide
(MBA) (at a given concentration). Immediately after addition of ammonium persulfate (0.5 ml of 5% solution), the
water phase was emulsified in the chilled organic phase
consisting of toluene (36 ml), chloroform (14 ml), TEMED
(0.08 ml) and Span 85 (at a given concentration) under a
nitrogen atmosphere. Emulsification was attained by stirring a bar (length=3.0 cm, width=0.7 cm) at a constant stirring speed which was checked stroboscopically. Polymerization started within a few minutes under a fluorescent light
and was completed after about 15 min. All solutions were

were kept in ice at 0-4°C. Polymerized acrylamide gel beads
were sometimes immediately used for activity determination
but usually washed thoroughly with the phosphate buffer and
lyophilized.

   Condition for preparation of beads. According to Dahlqvist et al (5), the catalyst system (TEMED and ammonium
persulfate) was added to the water phase. In our experiences, this method did not sometimes work well since the
water phase began to polymerize before it was emulsified
in the organic phase. In the present preparation, TEMED
was added to the organic phase and polyacrylamide gel beads
were satisfactorily obtained at any run.

   Since droplet size of the water phase in mechanically
prepared emulsions is not uniform, the diameter of the beads
varies over some ranges. Their mean diameters, however, can
be controlled by stirring speed and by concentration of the
emulsifying agent. These are shown in Figure 1. At higher
stirring speed and higher concentration of Span 85, diameters
tend to be smaller and more uniform. The condition for preparation of beads was set to 600 rpm and Span 85 concentration of 1.5% for diffusional effect within beads to be neglected.

   Standard determination of enzyme activity. The activity of the enzyme was determined at 30°C and shaking condition of 100 rpm towards the substrate lactose, the concentration of which was 4.5% (0.125 M) in M/15 phosphate
buffer at pH 6.5. After 30 minutes reaction, one ml of
reaction solution was taken in a pipet, the top of which
was covered with gauze to prevent contamination of enzyme
beads from the reaction mixture, and mixed with one ml of
10% trichloroacetic acid solution. The amounts of glucose
liberated were determined by glucose oxidase-catalase-
chromogen reagent.

   Diffusional effects. The enzymatic reaction in the
beads may be to some extent diffusion-controlled. In homogeneous aqueous solution, even the fastest enzymatic reaction appears not to be diffusion-controlled, but this is no
longer the case when the substrate has to diffuse toward
the enzyme through the stagnant layer between solution and
beads and within beads. Activity of the immobilized enzyme
was measured in a stirred tank reactor with baffles at 1000

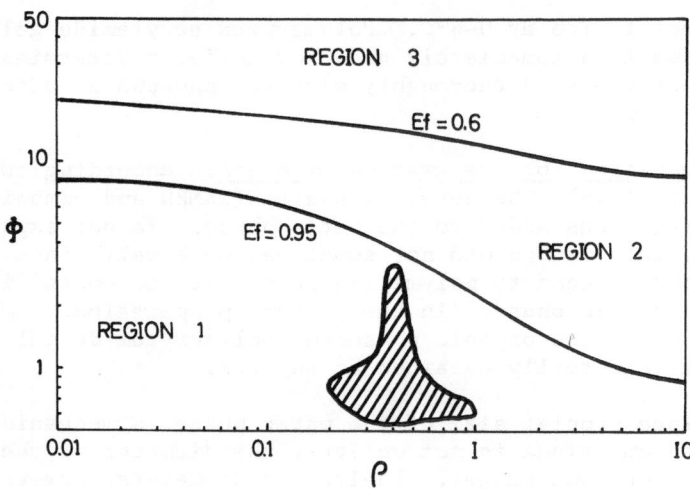

<u>Fig. 2.</u>: Plots of Φ against ρ for spherical particles. The
two lines correspond to the effectiveness factors
shown and divide the diagram into three regions
corresponding to : 1, small; 2, moderate and; 3,
large diffusional effects. All experimental data
are within the hatched region.
$\Phi = vR^2/D_S S$ and $\rho = K_m/S$ where v = overall reaction rate per unit volume of beads, R = radius of
beads, D_s = diffusivity of substrate within beads,
S = substrate concentration at surface of beads,
and K_m = Michaelis constant for immobilized enzyme.

rpm and was the same as that measured by the standard shaking method. Diffusional effect through the stagnant layer
is therefore negligibly small.

By applying the data of activity measurements and Michaelis constant mentioned below to the theory reported by
Kobayashi and Laidler (11), diffusional effect within beads
was estimated to be negligible (Figure 2). As both Φ and
ρ are obtained experimentally, they can be plotted in this
figure. Therefore, it is possible to know the region in
which experimental data fall. In the present case, they
fall in region 1 and intraparticle diffusional effect was
estimated to be negligibly small. Measured kinetic data

TABLE 1

Effect of Protective Agent During Polymerization on
Activity of Immobilized β-galactosidase

Protective Agent (18.7 mg/ml)	Relative Activity (%)		
	A. oryzae	K. lactis	E. coli
none	100	100	100
glucono-δ-lactone	165	27	6
galactono-γ-lactone	145	162	62
phenyl β-thiogalactoside	87	--	--
bovine serum albumin	142	78	110
casein	62	--	--
glutathione	162	2	6
dithiothreitol	134	142	145

are therefore free from both interparticle and intraparticle
diffusional effects.

Effect of MBA concentration on enzyme activity yield.
Effect of MBA on enzyme activity yield was measured by
varying concentration of MBA from 1.25% to 6.25%. Activity
yield was evaluated as :

Activity yield (%)

$$= \frac{\text{Activity in total enzyme beads}}{\text{Total activity of free enzyme used for immobilization}} \times 100$$

(1)

Activity yield of enzyme beads measured immediately after
preparation was the highest in the case of 1.25% MBA. However, the beads prepared with 1.25% MBA showed large activity loss during 2 hrs washing in the buffer, whereas those
with 2.5% MBA possessed the highest activity yield after
25 hrs washing and retained the same activity yield after
120 hrs washing. Concentration of MBA was therefore set to
2.5% in the subsequent experiments.

Effects of protective agents on activity yield. In
order to increase activity yield of the enzyme, some pro-

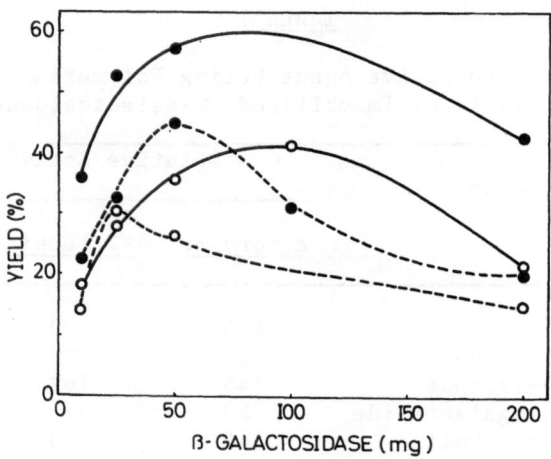

<u>Fig. 3</u>: Activity yield of β-galactosidase at various enzyme
            concentration.
            Solid line: the enzyme from K. lactis with (closed
            circle) and without (open circle) galactono-γ-
            lactone. Dotted line : the enzyme from A. oryzae
            with (closed circle) and without (open circle)
            glucono-δ -lactone.

tective agents were added to the enzyme-acrylamide-MBA
solution before polymerization at a constant amount of the
enzyme (50 mg). In the case of β-galactosidase from A.
oryzae, dithiothreitol and glutathione were examined in an
attempt to protect thiol groups of the enzyme against oxidation during polymerization and found to be effective as
shown in Table 1. Addition of bovine serum albumin also
increased the relative activity but addition of casein did
not. The presence of inhibitors, such as glucono-δ-lac-
tone and galactono-γ-lactone , during polymerization has
also effectively increased the relative activity, but phenyl
β-thiogalactoside revealed no advantageous effect.

     Protective effects during polymerization varied with the
origin of the enzyme. In the case of the enzyme from K.
lactis, galactono-δ -lactone was the best protective agent
and glucono-γ -lactone did not show protective effect. In
the case of the enzyme from E. coli, dithiothreitol was the

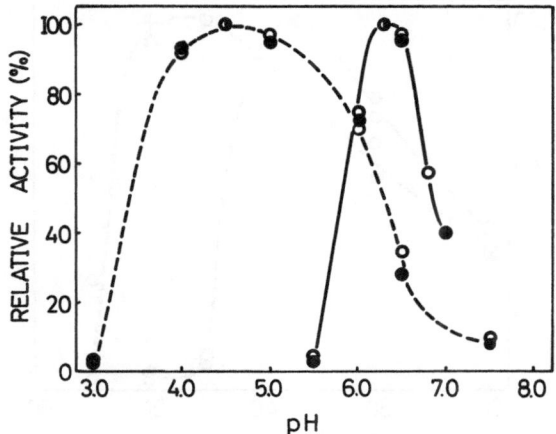

Fig. 4: Effect of pH on activity of β-galactosidase at 30°C.
         Solid line: the enzyme from K. lactis
         Dotted line: the enzyme from A. oryzae
         Open circle: immobilized enzyme
         Closed circle: free enzyme

best and both glucono- δ -lactone and galactono- γ -lactone
did not show protective effects.

Protective effects were also studied in variation of
amounts of the enzyme used for immobilization (Figure 3).
In the case of the enzyme from A. oryzae, activity yield of
the beads in the presence of glucono- δ -lactone increased
up to 46% even after thorough washing, whereas that without
the agent was only 28%. In the case of the enzyme from K.
lactis, maximum activity yield was almost 60% in the presence of galactono- γ -lactone, whereas that without the reagent was 42%.

Characteristics of immobilized enzyme. Optimum pH of
the immobilized enzyme from A. oryzae was about 4.5 in M/10
McIlvaine buffer and that from K. lactis was 6.3 in M/15
phosphate buffer as shown in Figure 4. Optimum temperature
of the immobilized enzyme from A. oryzae was 45°C and that
from K. lactis was 37°C as shown in Figure 5. All the relative values obtained from immobilized enzyme at various pH
and temperature were almost the same as those from free enzyme. Optimum pH and temperature of free enzyme from E.

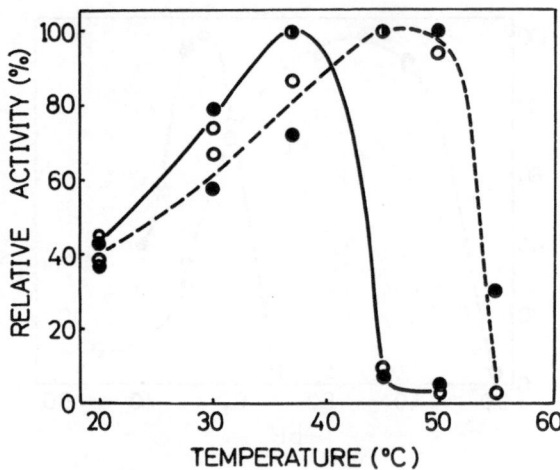

<u>Fig. 5</u>:   Effect of temperature on activity of β-galactosi-
            dase at pH 6.5 (M/15 phosphate buffer).
            Solid line: the enzyme from <u>K. lactis</u>
            Dotted line: the enzyme from <u>A. oryzae</u>
            Open circle: immobilized enzyme
            Closed circle: free enzyme

<u>coli</u> were 7.5 and 50°C, respectively, in M/15 phosphate
buffer. Those for immobilized enzyme from <u>E. coli</u> may be
the same.

        Thermostability of the immobilized enzyme was also si-
milar to that of free enzyme. The immobilized enzyme from
<u>A. oryzae</u> was fairly stable during 2 hrs incubation at 37°C
and pH 6.5, but the activity decreased remarkably at 45°C
and above. The activity decreased slightly at 30°C during
40 hrs holding but remarkably did at 37°C and above.

        In the case of immobilized enzyme from <u>K. lactis</u>, 20%
decrease in the activity was observed during 24 hrs incubation at 30°C. The activity, however, decreased very slightly
at 20°C.

        The Michaelis constants were evaluated from activity
measurements at various lactose concentrations. The constants for immobilized enzyme from <u>A. oryzae</u> were 46 mM at

TABLE 2

Kinetic Constants for β-galactosidase at 30°C and
pH 6.5 (M/15 phosphate buffer)

Origin	K_m(mM) for lactose	K_i (mM) for galactose	K_i (mM) for glucose
A. oryzae	46	1.2	#
K. lactis	34	46	ca. 2
E. coli	6.5	#	-

: so large that inhibition constant cannot be evaluated.

pH 6.5 and 20 mM at pH 4.5. Both values were almost the same as those of free enzyme. Table 2 shows kinetic constants for β-galactosidase at 30°C and pH 6.5. Inhibition of both products was competitive.

In many papers optimum temperature and thermostability of enzymes immobilized by the entrapping method were found to be almost similar to those of free enzymes, which were also the case in the present investigation. From the theory (12), it is expected that optimum pH and kinetic constants of immobilized enzymes are the same as those of free enzymes when support or matrix has no charge and there are freed from diffusional effects. Polyacrylamide beads have no charge and negligibly small diffusional effects were confirmed as described above. Thus, it is reasonable that measured optimum pH and kinetic constants of the immobilized enzyme are almost the same as those of free enzyme.

Lyophilized beads were stored at 4°C for a given time and activities of the beads were measured. No loss of enzymatic activity was measured after 47 days.

Hydrolysis of lactose in skim milk as well as that in lactose solution by the immobilized enzyme from A. oryzae was carried out in a shaken flask. Results are shown in Figure 6. The activity in skim milk was lower about 10% than that in lactose solution. Some factors present in skim milk, such as many kinds of protein, would interfere with hydrolysis of lactose.

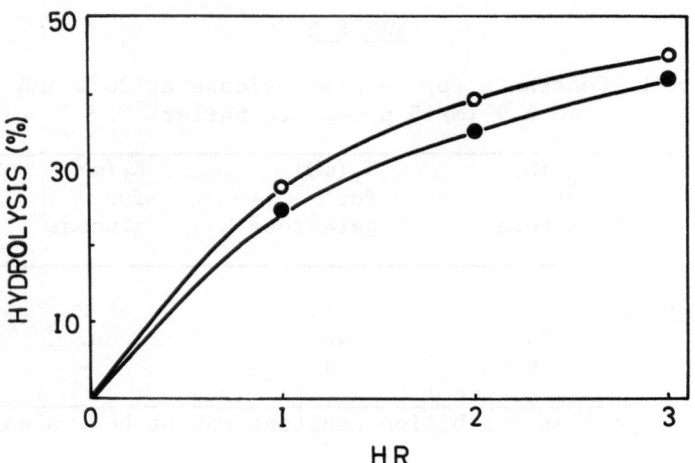

Fig. 6: Hydrolysis of lactose by immobilized enzyme from
          A. oryzae in the solution of skim milk (closed
          circle, initial lactose : 4.5%, pH 6.5) or in the
          phosphate buffer at pH 6.5 (open circle, initial
          lactose : 4.5%) in a shaken flask.

KINETICS OF IMMOBILIZED ENZYME IN CONTINUOUS TUBULAR REACTOR

In the case of the enzyme from A. oryzae, it was found
from initial reaction rate measurements that galactose reduced the rate of hydrolysis of lactose competitively, whereas glucose did not (Table 2). Thus, the reaction rate, v,
during the course of hydrolysis of lactose may be expressed
as

$$v = \frac{V_m S}{K_m \left(1 + \frac{I}{K_i} \right) + S} \quad , \quad I = S_o - S \qquad (2)$$

where S_o = initial lactose concentration, K_i = inhibition
constant for galactose, V_m = maximum reaction rate, S =
substrate concentration, and I = concentration of inhibitor.

For a batch reactor assuming no product present initially, equation (2) is integrated as

$$K_m(1 + \frac{S_o}{K_i}) \ln \frac{S_o}{K_i} - (\frac{K_m}{K_i} - 1) (S_o - S) = V_m t \quad (3)$$

where t = reaction time. In the case of S_o = 125 mM which corresponds to the average lactose concentration of bovine milk in Japan, time course of enzymatic hydrolysis of lactose at pH 6.5 was found to be expressed by equation (3) up to at least 80% of hydrolysis.

To develop kinetics of immobilized enzyme in continuous tubulor reactor, certain general assumptions, which have been found to be valid in many cases, are made. They are as follows: (a) Reactor is maintained at isothermal condition, (b) Flow in a reactor is represented by plug-flow and no backmixing effect due to longitudinal turbulent dispersion, (c) There are no interparticle and intraparticle diffusion limitations, (d) Enzyme beads are packed uniformly in a reactor. By substituting equation (2) in material balance equation of an idealized reactor at steady state, the following equation is derived:

$$K_m(1 + \frac{S_o}{K_i}) \ln \frac{1}{1 - x} - (\frac{K_m}{K_i} - 1) S_o x = \frac{V_m (1 - \epsilon)}{S_v} \quad (4)$$

where x = degree of hydrolysis in the reactor outlet, S_v = space velocity (= volumetric flow rate of substrate divided by volume of the reactor) and ϵ = void fraction of the reactor.

Relationships between the degree of hydrolysis and the flow rate at various initial lactose concentration were obtained in a continuous tubular reactor under the condition that the degree of hydrolysis is less than 80%. By applying these data to equation (4), kinetics of immobilized enzyme in tubular reactor was found to follow the performance equation (4) when the superficial velocity of substrate solution is greater than 3 cm/min. At low superficial velocity than 3 cm/min, the apparent Michaelis constant increased as decreasing superficial velocity, which was similar result as that reported by Kobayashi and Moo-Young (13). This is due to the interparticle diffusion effect existing between surface of beads and bulk solution. However, at high superficial velocity than 3 cm/min, the interparticle diffusion effect becomes negligible and the apparent Michaelis constant was independent of flow rate; justification of assumption (c) was made clear.

As for assumption (b), theoretical results (14) were applied in the present case. Immobilized enzyme reactors in laboratory scale have sometimes the possibility of significant backmixing effect on the degree of hydrolysis. Even in industrial scale operation, backmixing effect may become substantial if large particles are used. According to the axial dispersion model, the material balance for the substrate gives the following differential equation:

$$D_z \frac{d^2S}{dz^2} - u \frac{dS}{dz} - \frac{k_1 a}{\epsilon} (S - S_s) = 0 \tag{5}$$

with the boundary conditions:

$$S - \frac{D_z}{u} \frac{dS}{dz} = S_o \qquad \text{at } z = 0 \tag{6}$$

$$dS/dz = 0 \qquad \text{at } z = L \tag{7}$$

where D_z = axial dispersion coefficient, z = axial distance of reactor, u = superficial velocity, k_1 = mass transfer coefficient in liquid film, a = external surface area of beads per unit volume of reactor, S_s = concentration of lactose at the surface of beads, and L = reactor length. By solving above equations, the effect of backmixing on the degree of lactose was estimated to be negligibly small.

As assumptions (a) and (d) are valid in most cases and assumptions (b) and (c) were checked to be valid at high superficial velocity, feasibility of reactor scale-up was indicated.

Hydrolysis of lactose in skim milk as well as that in lactose solution by the immobilized enzyme from A. oryzae was done in a continuous tubular reactor. Results are shown in Figure 7. As was observed in shaken flask experiments, about 10% decrease of the activity in skim milk was also observed as compared with that in lactose solution.

Operational stability of the immobilized enzyme from A. oryzae was studied in a continuous tubular reactor at

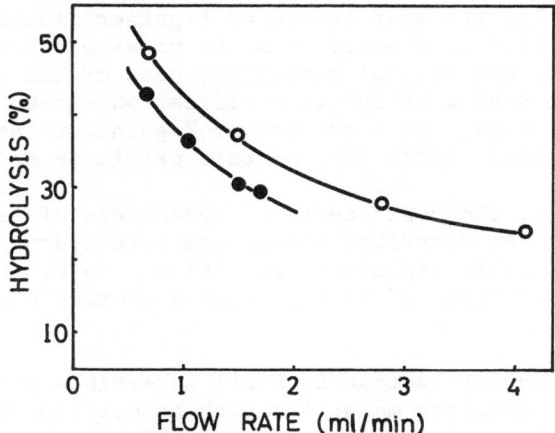

<u>Fig. 7</u>:   Hydrolysis of lactose by immobilized enzyme from
          <u>A. oryzae</u> in the solution of skim milk (closed
          circle, initial lactose : 4.5%, pH 6.5) or in the
          phosphate buffer at pH 6.5 (open circle, initial
          lactose : 4.5%) in continuous tubular reactor.
          Diameter of reactor : 1.0 cm.

30°C and pH 6.5. Decrease of the immobilized enzyme activity on the whole column was about 16% after 48 hrs. operation.

      The present investigation demonstrates the potential
for the preparation of low lactose milk. However, to develop economical process to produce low lactose milk, there
are many problems to be solved. For example, partially purified β-galactosidase was used for immobilization in the
present investigation. Thus, polyacrylamide gel beads may
contain other kinds of enzyme, and other substances besides
galactose and glucose may be produced. It was found from
the results of thin layer chromatography for lactose and
skim milk solution treated with enzyme beads that both solutions certainly contain unknown carbohydrates, although, except the sweetness, no difference between treated and un-
treated milk was recognized by the sensory test. These carbohydrates must be identified and their safety to drink must
be checked.

Secondly, the best origin of β-galactosidase for immobilization must be checked from the view-points of optimum temperature, operational stability, cost of the enzyme including the degree of enzyme purification and so on.   In the present work, three origins of β-galactosidase were studied. Wider search for the best origin is necessary.

Thirdly, the best degree of hydrolysis of lactose must be decided. As described above, complete hydrolysis of lactose is desirable especially for Asian people.   However, complete hydrolysis of lactose causes another problem (15, 16).

Existance of lactose in small intestines promotes absorption of bivalent metal ion such as calcium or ferrous ion.   If infants drink everyday the milk of which lactose is completely hydrolyzed, there may happen anemia due to the shortage of ferrous ion. Shortage of calcium ion may cause retardation of development of bones. Thus, the best degree of hydrolysis of lactose must be decided. No existance of acute and chronic toxic substances must also be checked.

Acknowledgement. The authors wish to express their sincere gratitude to Tokyo Tanabe Co., Ltd. for providing β-galactosidase from A. oryzae.

REFERENCES

1. Bayless, T.M. and Huang, S., Amer. J. Clin. Nutr., 22, 250 (1969).

2. Kretchmer, N., Sci. Amer., 227, No. 4, 70 (1972).

3. Sasaki, Y., Ito, M., Kameda, H., Ueda, H., Aoyagi, T., Christopher, N.L., Bayless T.M. and Wagner, H.N., J. Lab. Clin. Med., 76, 824 (1970).

4. Herbst, J.J., Sunshine, P. and Kretchmer, N., Advances in Pediatrics, 16, 11, Year Book (1969).

5. Dahlqvist, A., Mattiasson, B. and Mosbach, M., Biotechnol. Bioeng., 15, 395 (1973).

6. Hustad, G.O., Richardson, T. and Olson, N.F., J. Dairy Sci., 56, 1111 and 1118 (1973).

7. Woychik, J.H. and Wondolowski, M.V., J. Milk Food Tech-
    nol., 36, 31 (1973).

8. Okos, E.S. and Harper, W.J., J. Food Sci., 39, 88 (1974).

9. Weetall, H.H., Havewala, N.B., Pitcher, W.H., Detar, C.
    C., Vann, W.P. and Yaverbaum, S., Biotechnol. Bioeng.,
    16, 295 and 689 (1974).

10. Wierzbicki, L.E., Edwards, V.H. and Kosikowski, F.V.,
    Biotechnol. Bioeng., 16, 397 (1974).

11. Kobayashi, T. and Laidler, K.J., Biochim. Biophys. Acta,
    302, 1 (1973).

12. Kobayashi, T. and Laidler, K.J., Biotechnol. Bioeng.,
    16, 77 (1974).

13. Kobayashi, T. and Moo-Young, M., Biotechnol. Bioeng.
    15, 47 (1973).

14. Kobayashi, T. and Moo-Young, M., Biotechnol. Bioeng.,
    13, 893 (1971).

15. Hegsted, D.M., "Present Knowledge in Nutrition", 3rd
    Ed., The Nut. Foundation Inc., p. 147 (1967).

16. Ebihara, K. and Yoshida, A., Abstract of Annual Meeting
    of the Japanese Society of Food and Nutrition (1974).

IMMOBILIZATION OF ENZYMES BY RADIATION POLYMERIZATION

Hideo Suzuki, Hidekatsu Maeda and Aizo Yamauchi[*]

Fermentation Research Institute; [*]Research Institute for Polymers and Textiles - Inagehigashi, Chiba City; [*]Kanagawa, Yokohama City, Japan

INTRODUCTION

Many studies have been reported on the immobilization of enzymes. One of the immobilizing techniques which have been employed is the gel entrapping method. Bernfeld reported that the enzyme can be entrapped within the matrices of the gel network by chemically polymerizing and cross-linking acrylamide monomer and N,N'-methylenebisacrylamide (1).

The authors have investigated the use of radiation polymerization in place of chemical polymerization in the gel entrapping method and have found that the enzyme can be entrapped within the gel matrices resulting from the radiation polymerization of synthetic monomers or polymers.

In this paper, the authors wish to report on the production of immobilized enzyme preparations by radiation polymerization and some general properties of the preparations obtained.

MATERIALS AND METHODS

Enzymes. Glucoamylase of Asp. niger, a product of Miles Laboratory; invertase of Candida utilis, purchased from Seikagaku Kogyo Co., Ltd.; -galactosidase of Asp. oryzae, a product of Tokyo Tanabe Pharmacy Co., Ltd. Those enzymes were used in this work after dialysis against tap water.

Preparation of Immobilized Enzyme in Synthetic Monomer
Enzyme System. Five grams of synthetic monomer was dissolve
in 20 ml of distilled water or buffer solution saturated
with nitrogen gas and 2.5 ml of enzyme solution was added
to the solution with stirring under a nitrogen atmosphere.
A 5 ml portion of the mixture was poured into an ampoule.
The ampoule was sealed in a nitrogen atmosphere and irrad-
iated with γ -ray at a rate of 6.54×10^4 or 10×10^4 rad/hr.
After the irradiation, the gel formed was picked out from
the ampoule and homogenized by a homogenizer. The homogenized gel was washed 3 times with distilled water and weighed
after draining.

Preparation of Immobilized Enzyme in Synthetic Polymer-
Enzyme System. Ten grams of polyvinylalcohol was suspended
in 90 ml of distilled water saturated with nitrogen gas and
thoroughly dissolved by heating the solution to 80-90°C.
After the solution had cooled in an atmosphere of nitrogen,
10 ml of enzyme solution was added with stirring under a
nitrogen atmosphere. A definite amount of the mixture,
generally 4-6 ml, was poured into an ampoule. The ampoule
was sealed in a nitrogen atmosphere and irradiated with γ -
ray as previously described. After irradiation, the gel was
picked out from the ampoule and chopped with a razor. The
chopped gel was washed 3 times with distilled water and
weighed after draining. Equipment manufactured by Tokyo
Shibaura Electric Co., having 500 Ci of ^{60}Co as the irradiation source was used in this work.

Preparation of Immobilized Enzymes by Electron-Beam
Irradiation. A definite amount of the mixture of enzyme
and polyvinylalcohol, generally 4-6 ml, was poured into a
flat ampoule (18x220 mm; thickness of ampoule, 5.8 mm and
thickness of glass, 1.0 mm). The ampoule was sealed in a
nitrogen atmosphere and irradiated at a rate of 1.2 mega
roentgen/min. It was shaken over a width of 200 mm at a
speed of 10 mm/sec under the electron-beam while being
cooled. An electron-beam of 7.5 million ev and 49.5 A
generated by the linear accelerator of the Government Industrial Research Institute, Nagoya, was used in this work.

Estimation of Enzyme Activity. The reaction mixture
used in this work was as follows: 5ml of 2% sugar solution,
4 ml of buffer solution and 1 ml of enzyme suspension. Maltose, lactose and sucrose were the substrates of the glucoamylase, β-galactosidase and invertase, respectively. The

reaction mixture was incubated at 40°C for 30 min under a
shaking of 130 rev/min. After the reaction, the mixture
was maintained in a boiling water bath for 10 min to stop
the enzyme reaction and the glucose liberated by the enzyme
reaction was estimated by a Glucostat. One unit was defined as the activity liberating 1 μ mole of glucose per
minute under the given conditions.

Test for Leaking of the Entrapped Enzyme. This test
was carried out as follows: 1 or 2 g of homogenized gel
was suspended in a mixture of 20 ml of substrate solution
and 20 ml of buffer solution. The suspension was incubated
for 60 min at 40°C with shaking. After the incubation, the
suspension was filtered to remove the gel and the enzyme
activity in the filtrate was estimated after dialysis against
tap water. It was found that enzyme gel which registered
negative in this leaking test could hydrolyze substrate continuously in a column system for at least 2 weeks maintaining a constant hydrolysis ratio.

RESULTS AND DISCUSSION

Preparation of Immobilized Enzyme in Acrylamide-Enzyme
System. The enzyme gels were prepared by subjecting the
acrylamide-enzyme system to γ-ray radiation (2). As enzymes, glucoamylase, invertase and β-galactosidase were
used separately. The results were as shown in Table I.
The entrapped activity was defined as the activity of the
supplied enzyme in this system minus the activity found in
the distilled water used in washing the homogenized gel.
The relative activity of the gel against the native activity
was represented as B/A. This system gelled at the irradiation of 1 mega rad and the preparations did not leak the
entrapped enzymes except in the case of glucoamylase gel.
Glucoamylase gel without leakage could, however, be prepared under irradiation of more than 2 mega rad. The weight
of the gel decreased with the increase of radiation dose.
This shows that increased radiation dose allows formation
of gel having more rigid structure. The realtive activities
of glucoamylase gel, invertase gel and β-galactosidase gel
without leakage were 40%, 25% and 33%, respectively.

Preparation of Immobilized Enzyme in Dimethylacrylamide-
Enzyme System. Enzyme gels having higher relative activities
were prepared in dimethylacrylamide-enzyme system (3). Typical results obtained are shown in Table 2. The relative ac-

TABLE 1

Enzyme	Rad	Activity (U)			B/A	Leakage
		Added	Entrapped(A)	Gel(B)		
Glucoamylase	1×10^6	108.4	99.6	40.9	41	++
	2×10^6	108.4	102.8	40.8	40	-
	4×10^6	108.4	108.4	16.9	16	-
Invertase	1×10^6	2021	1909	482	25	-
	2×10^6	2021	1999	442	22	-
	4×10^6	2021	2015	145	7	-
-galactosidase	1×10^6	130.1	125.8	40.9	33	-
	2×10^6	130.1	130.1	25.8	20	-
	4×10^6	130.1	130.1	0	0	-

TABLE 2

Enzyme	Rad	Activity (U)			B/A	Leakage
		Added	Entrapped(A)	Gel(B)		
Glucoamylase	2×10^6	123	113.1	59.5	53	-
Invertase	2×10^6	4490	4324	1230	28	-
-galactosidase	2×10^6	81.8	78.4	43.9	56	-

TABLE 3

| Enzyme | Rad. | Activity (U) | | | | Leak-age |
		Added	Entrapped(A)	Gel(B)	B/A	
Glucoamylase	1.1×10^6	93.1	57.6	28.5	50	+
	2.1×10^6	93.1	67.9	27.0	40	±
	3.8×10^6	93.1	80.8	25.4	31	±
	5.0×10^6	93.1	84.6	21.3	25	±
	6.4×10^6	93.1	87.8	15.7	18	−

tivities of glucoamylase gel, invertase gel and β-galactosidase gel were 53%, 28% and 56%, respectively.

Preparation of Immobilized Enzyme with Other Acrylic Monomers. The gelation profile of 2-hydroxyethylacrylate-enzyme system was rather unusual (3). This system gelled with a low radiation dose on the order of 0.25 mega rad but the resulting preparations leaked the entrapped enzyme. Gells without leakage were obtained by an irradiation dose of 5-6 mega rad. In the preparation of glucoamylase gel, an irradiation dose of 6.4 mega rad made it possible to prepare a gel without leakage as shown in Table 3, but the relative activity of the gel obtained at that time was as low as 18%. In the case of sodium acrylate-enzyme system, the gelation did not occur even at the irradiation of 4 mega rad (3).

TABLE 4

| Enzyme | Rad. | Activity (U) | | | | Leak-age |
		Added	Entrapped(A)	Gel(B)	B/A	
Glucoamylase	2.9×10^6	92.1	92.1	41.7	45	−
	3.9×10^6	92.1	92.1	35.0	38	−
	5.9×10^6	92.1	92.1	21.8	24	−
β-Galactosidase	2.9×10^6	170	170	14.6	8.6	−
	3.9×10^6	170	170	12.6	7.4	−
	5.9×10^6	170	170	5.9	3.5	−
Invertase	4.0×10^6	974	938	9.9	1.1	−
	6.0×10^6	974	963	4.3	0.4	−

TABLE 5

Enzyme	Rad	Activity (U) Entrapped(A)	Gel(B)	B/A	Leakage
Glucoamylase	4×10^6	139	22.5	16	±
	5×10^6	110	26.2	24	-
	6×10^6	112	22.3	20	-
	7×10^6	124	21.7	18	-
Invertase	4×10^6	8778	1466	17	++
	5×10^6	5947	1084	18	-
	6×10^6	11210	828	7	-
	7×10^6	8360	783	9	-

Preparation of Immobilized Enzyme in N-Vinylpyrrolidone-Enzyme System. The results obtained with preparation
of immobilized enzyme in N-vinylpyrrolidone-enzyme system
are shown in Table 4. The gelation occurred at the irradiation of 1 mega rad but 2.9 mega rad was required to obtain
enzyme gel without leakage. The relative activities of enzyme gels were much influenced by the kind of enzyme (4).
In glucoamylase gel, the relative activity was 45%, but in
β-galactosidase gel, it was only 8.6%. In invertase gel
prepared at 4 mega rad, the relative activity was as low as
1%. These results suggest that in the N-vinylpyrrolidone-
enzyme system, the relative activities of enzyme gels depend on the kind of enzyme. However, the reason has not
yet been clarified.

Preparation of Immobilized Enzyme in Synthetic Polymer-
Enzyme System. As a synthetic polymer, polyvinylalcohol was
used in this work. The polyvinylalcohol-enzyme system gelle

TABLE 6

Enzyme	Rad	Activity (U) Entrapped(A)	Gel(B)	B/A	Leakage
Glucoamylase	6×10^6	53.2	26.2	49	-
	12×10^6	54.8	19.1	35	-
	36×10^6	72.9	1.8	2.4	-
Invertase	6×10^6	782	342	44	-
	12×10^6	611	179	29	-
	36×10^6	934	4	0.4	-

TABLE 7

Enzyme Conc.(%)	Added	Activity (U) Entrapped(A)	Gel(B)	B/A	Leakage
0.26	80.1	77.2	49.5	65	-
0.52	160	146	95.9	66	-
1.04	320	293	1,80	61	-
2.08	640	481	335	70	±
4.16	1280	1070	587	55	±

with a radiation dose of more than 4 mega rad and a preparation without leakage was obtained with a radiation dose of
more than 5 mega rad as shown in Table 5. The relative
activities of glucoamylase gel and invertase gel prepared
at the irradiation dose of 5 mega rad were 24% and 18%,
respectively (5). In this case, the entrapped activity
was calculated from the amount of ammonia which resulted
from Kjeldahl's digestion of enzyme gel.

Preparation of Immobilized Enzyme by Electron-Beam
Irradiation. The authors tried using electron-beam, instead of γ-ray, for the preparation of enzyme gel with polyvinylalcohol-enzyme system, because this system requires
high radiation dose on the order of 4 mega rad for the gelation. The results obtained are shown in Table 6. The relative activities of glucoamylase gel and invertase gel
prepared at the irradiation dose of 6 mega roentgen were
49% and 44%, respectively (6). As mentioned above, when
γ-ray was applied to this system, the relative activities
at glucoamylase gel and invertase gel were 24% and 18%, respectively. Thus, preparations possessing higher relative
activities were obtained by electron-beam irradiation.
With this linear accelerator, the time needed to generate
6 mega roentgen was only 5 min. On the other hand, about
90 hours in γ-ray irradiation are required to produce the
same dose because the equipment used in this work contains
only 500 Ci of ^{60}Co. It is still unclear whether the
superiority of electron-beam is due to the shortening of
irradiation period or due to the kind of radiation ray.

The Effect of Enzyme Concentration on the Gel Formation
The effects of enzyme concentration on the gelation were
studied in acrylamide-glucoamylase system and dimethylacrylamide-glucoamylase system (3). The concentration of glucoamylase in the acrylamide-glucoamylase system was varied in

TABLE 8

	Activity (U)			
	Control	1 M rad	2 M rad	4 M rad
Glucoamylase	329.3	321.4	318.4	269.5
Invertase	7547	6248	5887	4502
β-galactosidase	371.3	316.2	284.5	227.3

the range of from 0.26% to 4.16% and the systems were irradiated by γ-ray at 3 mega rad. As shown in Table 7,
glucoamylase gels without leakage were obtained with the
systems containing glucoamylase below 1% and the gels prepared with the systems containing the enzyme at more than
2% leaked the entrapped enzyme. The same tendency was observed in dimethylacrylamide-glucoamylase system. It is
necessary to maintain enzyme concentration in the system
below 1% for the preparation of enzyme gel without leakage.

    Inactivation of Enzyme by Radiation. Aqueous solution
of the enzymes was irradiated by γ-ray and the results obtained are shown in Table 8. Glucoamylase still had about
80% of its original activity after irradiation of 4 mega
rad. Under the same conditions, the activity of invertase
was reduced to about 60% of its original value and the activity of β-galactosidase was reduced to about 70% (2). In
our preparation at least 2 mega rad is required, and therefore, there is a limitation on the kind of enzyme to be entrapped. Enzymes sensitive to radiation rays such as urease,
cannot be used in our preparation. As the toxicity of polyvinylalcohol is far less than that of acrylamide, more hope
can be held for the application of enzyme gel prepared with
polyvinylalcohol-enzyme system to the food industry.

TABLE 9

	Activity (U)		
Dextrin (M.W.)	Entrapped(A)	Gel (B)	B/A
16,000	729	165	23
6,500	666	276	41
4,750	686	222	32
4,000	691	339	49
3,000	713	353	50
1,900	729	367	50
1,100	691	368	51

General Properties of Glucoamylase Gel. The authors
studied the general properties of the glucoamylase gel
which was prepared by irradiating N-vinylpyrrolidone-glucoamylase system at 2.9 mega rad (4). The particle size of
the gel used in this experiment was in the range of 42-250
mesh. The optimum pH of glucoamylase gel was almost the
same as that of native glucoamylase. The heat stability of
glucoamylase gel was slightly inferior to that of native
enzyme as shown in Fig. 1. As shown in Fig. 2, the pH
stability of the gel was almost identical to that of native
enzyme in acidic and neutral range but in alkaline range it
was rather inferior to that of native enzyme. The Km of

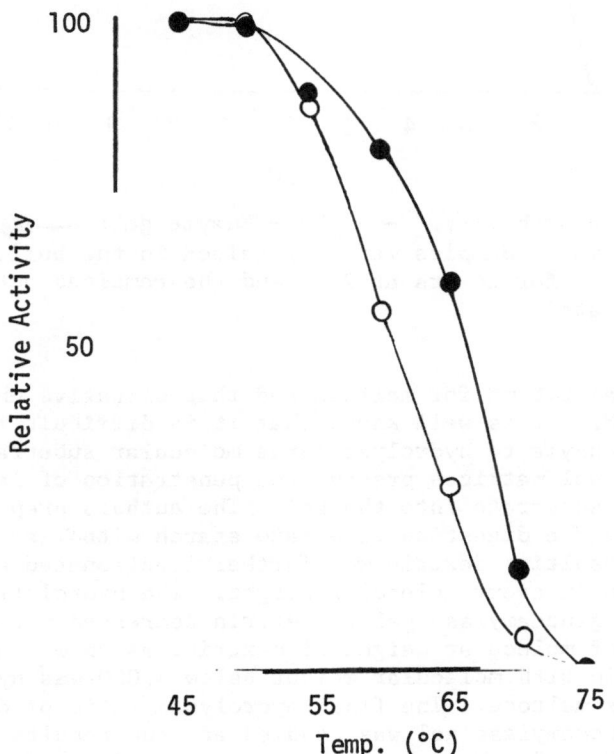

Fig. 1: Heat Stability ——O—— Enzyme gel : ——●——
Native Enzyme. Samples were maintained at the appointed
temperature for 15 min and the remained activities were
estimated.

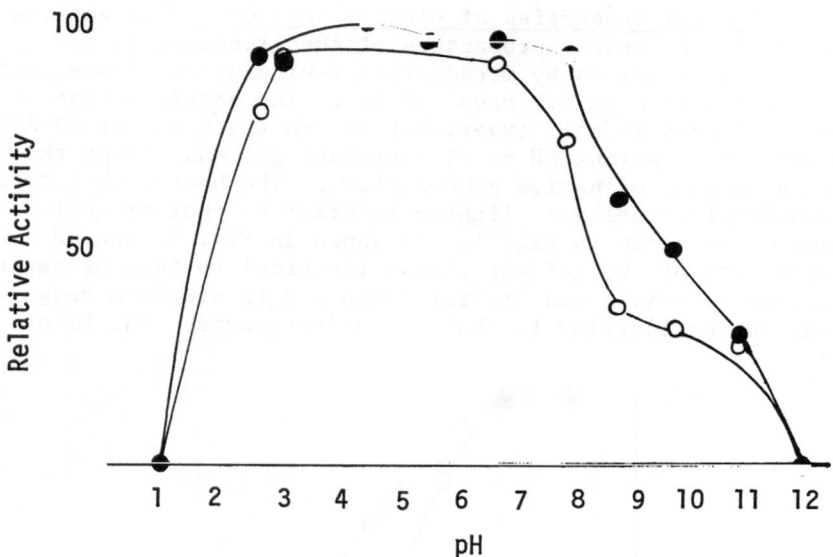

Fig. 2: pH Stability. ——O—— Enzyme gel: ——●——
Native enzyme. Samples were maintained in the buffer of
appointed pH for 24 hrs at 25°C and the remained activities
were estimated.

the gel was 2.8 mM for maltose and that of native enzyme
was 1.1 mM. It is well known that it is difficult for entrapping enzyme to hydrolyze large molecular substrate because the gel matrices prevent the penetration of large
molecular substrate into the gel. The authors prepared
dextrin by the digestion of potato starch with α-amylase
and the resulting dextrin was further fractionated by gel
filtration in every molecular weight. The hydrolytic activities of glucoamylase gel to dextrin decreased with the
increase of molecular weight of dextrin, as shown in Table
9. Dextrin with molecular weight below 4,000 was hydrolyzed
as well as maltose. The final hydrolysis ratio of dextrin
by the glucoamylase gel was studied and the results are
shown in Fig. 3. Dextrin with molecular weight of 1,100
was hydrolyzed to glucose with the hydrolysis ratio of 97%
by the glucoamylase gel after 4 hours' incubation. Dextrin
with molecular weight of 10,400 was also hydrolyzed by the
glucoamylase gel at the same ratio after 23 hours whereas

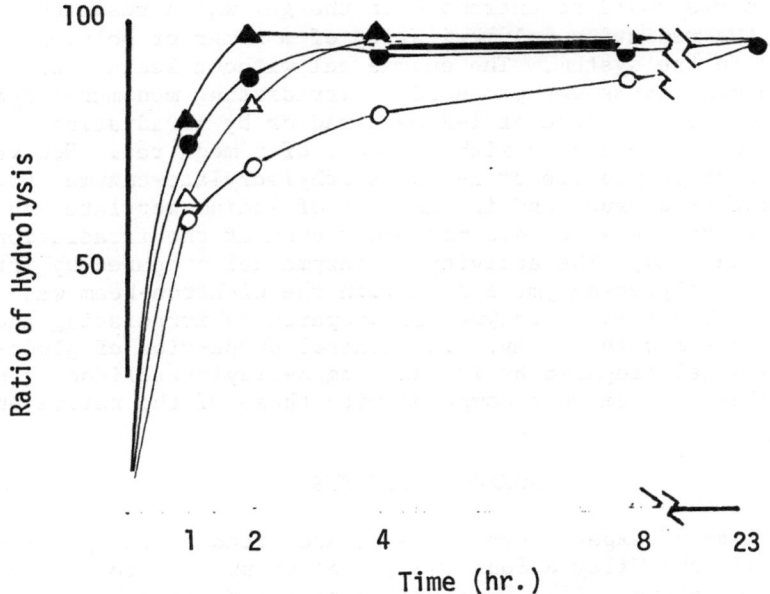

Fig. 3: Hydrolysis of Dextrin. ——O—— Dextrin (M.W. 10,400)-gel : ——●—— Dextrin (M.W. 10,400)-native —△—— Dextrin (M.W. 1,100)-gel : ——▲—— Dextrin (M.W. 1,100)-native. Incubation was carried out at 40°C.

with the native enzyme it takes about 8 hours. It is a very interesting phenomenon that dextrin with molecular weight of about 10,000 can be hydrolyzed by the glucoamylase entrapped in the gel.

SUMMARY

A new preparation of immobilized enzyme by the use of radiation polymerization was studied. Synthetic monomers such as acrylamide, dimethylacrylamide, 2-hydroxyethylacrylate, sodium acrylate and N-vinylpyrrolidone, and synthetic polymer such as polyvinylalcohol were used in this work. Glucoamylase, invertase and β-galactosidase were used as model enzymes. By subjecting a synthetic monomer-enzyme system or a polyvinylalcohol-enzyme system to radiation ray, the enzyme could be entrapped in the gel which resulted from the radiation polymerization of monomer or polymer added in the system. The enzyme gel without leakage of entrapped enzyme was produced by irradiating monomer-enzyme system with the dose of 1-2 mega rad or by irradiating polymer-enzyme system with the dose of 5 mega rad. However, the gelation profile of 2-hydroxyethylacrylate-enzyme system was rather unusual and in the case of sodium acrylate-enzyme system, the gelation did not occur even at the irradiation of 4 mega rad. The activity of enzyme gel prepared by irradiating polymer-enzyme system with the electron-beam was higher than that of enzyme gel prepared by irradiating the same system with γ-ray. The general properties of glucoamylase gel prepared by irradiating N-vinylpyrrolidone-glucoamylase system were compared with those of the native enzyme.

ACKNOWLEDGEMENTS

Some of experimental results are cited by the permission of John Wiley & Sons Inc., Publishers from the following literature. (1) "Preparation of Immobilized Enzymes Using Poly (vinyl Alcohol)" by Hidekatsu Maeda, Hideo Suzuki and Aizo Yamauchi; Biotechnol. Bioeng. 15 607. Copyright 1973 by John Wiley & Sons, Inc.
(2) "Preparation of Immobilized Enzymes by Electron Beam Irradiation" by Hidekatsu Maeda, Hideo Suzuki and Aizo Yamauchi, Biotechnol. Bioeng. 15, 827. Copyright 1973 by John Wiley & Sons, Inc.
(3) "Preparation of Immobilized Enzymes by N-Vinylpyrrolidone and the General Properties of the Glucoamylase Gel" by Hidekatsu Maeda, Hideo Suzuki and Aizo Yamauchi; Biotechnol. Bioeng. 16 in press. Copyright 1974 by John Wiley & Sons, Inc.
(4) "Preparation of Immobilized Enzymes by Acrylic Monomers under γ-ray Irradiation" by Hidekatsu Maeda, Hideo Suzuki and Aizo Yamauchi; Biotechnol. Bioeng. 17 in press. Copyright 1975 by John Wiley & Sons, Inc.

REFERENCES

1. P. Bernfeld and J. Wan, Science 142, 678 (1963).

2. H. Maeda, A. Yamauchi and H. Suzuki, Biochim. Biophys. Acta 315, 18 (1973).

3. H. Maeda, H. Suzuki and A. Yamauchi, Biotechnol. Bioeng. 17, in press.

4. H. Maeda, H. Suzuki and A. Yamauchi, Biotechnol. Bioeng. 16, in press.

5. H. Maeda, H. Suzuki and A. Yamauchi, Biotechnol. Bioeng. 15, 607 (1973).

6. H. Maeda, H. Suzuki and A. Yamauchi, Biotechnol. Bioeng. 15, 827 (1973).

REMOVAL OF HEAVY METAL ENZYME INHIBITORS

R.P. Chambers, G.A. Swan, E.M. Walle, W. Cohen
and W.H. Baricos
Department of Chemical Engineering & Biochemistry
Tulane University
New Orleans, Louisiana   70118

ABSTRACT

Protection of immobilized enzyme activity by removal of heavy metal enzyme inhibitors from enzyme reactor feed streams has been studied with a recently developed heavy metal selective adsorbent. Enzyme protection from mercury-caused inhibition was demonstrated for the enzymes glucose isomerase, alcohol dehydrogenase and the lactases from Aspergillus niger and Saccharomyces fragilis. The adsorbent, which contains thiol groups covalently bound to a porous inorganic matrix, forms tight complexes with mercury. Mathematical simulation and experimental confirmation of adsorbent kinetics and column performance indicated that mercury adsorption on the adsorbent proceeds by a "progressive shell" mechanism. The simulation was used to project adsorbent performance for large-scale enzyme protection.

The significance of enzyme deactivation resulting from the heavy-metal contamination of enzyme reactor feed streams was presented. Mercuric chloride was found to be a titrating inhibitor of glucose isomerase, alcohol dehydrogenase and Saccharomyces fragilis lactase. Activity loss as a function of the amount of mercuric chloride inhibitor was found to be different for soluble and immobilized glucose isomerase; at low inhibitor concentrations the immobilized enzyme was more severely inactivated by a given quantity of inhibitor, while at higher inhibitor concentrations, above that at which the soluble enzyme was completely inactivated, residual enzyme activity persisted.

199

A severe loss of enzyme activity may result from heavy metal enzyme inhibitors when they are present at trace levels in industrial immobilized enzyme reactors. Significant enzyme deactivation may result from long term flow of substrate solutions past the immobilized enzyme at concentrations sufficiently low that heavy metals may remain undetected by conventional atomic absorption analysis (1). Heavy metal inhibition can occur at the same time as other causes of enzyme inactivation such as enzyme denaturation, bacterial inactivation, and desorption of enzyme from the carrier matrix.

It is instructive to examine the magnitude of "background levels"of heavy metals which might be encountered in enzyme processes. Environmental studies have reported heavy metal concentrations at the part per billion (ppb) and sub-part per billion level in a number of aqueous environments (see Table 1). A larger quantity of heavy metal inhibitors in industrial process streams may arise from low level heavy metal contamination of raw materials. As an example of low level material contamaination, consider that a ten per cent solution of reagent grade NaCl (prepared in absolutely pure water) is estimated to contain about 100 ppb total heavy metals due to the impurities in that reagent grade chemical. Other important sources of heavy metals are often process-derived; trace concentration of heavy metals in the effluent from ion exchange columns, contaminants arising from upstream processes, piping or equipment, etc. Thus, heavy metals in the part per billion region can be expected in "highly purified" industrial enzyme reactor feed streams and higher concentrations could often be encountered. Additionally, besides the persistive low level contamination, one might anticipate "upsets" where the heavy metal concentration may be excessive for short time periods.

Inhibition by Heavy Metals

Heavy metals may bind with a number of functional groups present in enzymes and inactivate them (8). For example, they can bind to sulfhydryl groups, often an essential component of the active site of enzymes (9), or onto sites which are essential for maintaining structural conformation (10). Of particular importance is the magnitude of the binding constant and the degree of activity re-

<u>TABLE 1</u>

BACKGROUND LEVELS OF HEAVY METALS

Metal	Environments	Concentration ppb	Reference
Mercury	Rainwater (typical)	0.2	2
	Ground water (typical)	0.05	2
	LeHave River, Nova Scotia	0.04-0.2	25
Lead	Food Processing Plant Water	10-20	3
	City Water (avg. 100 largest cities U.S.)	3.7	4
	Drinking water (Max. limit)	50	5
Copper	Calif. Surface Water (Mean of 65)	18	6
Cadmium	Surface Water	< 1	7
Cobalt	Calif. Surface Water (Mean of 65)	4.3	6

storation possible upon solution replacement. Inhibitors that bind on the enzymes with very high binding constants (essentially irreversible equilibrium) and in which activity of the inhibited enzyme cannot be restored upon solution replacement will be referred to as "titrating inhibitors" in this paper. Those inhibitors with lower binding constant and in which the inhibited enzyme activity can be restored by placing the inhibited enzyme in a solution free of heavy metals (solution replacement) will be referred to as "self-reversing inhibitors".

The action of titrating inhibitors is particularly injurious to continuous processing enzyme reactors due to the progressive accumulation of inhibiting metal on the enzyme and the consequent cumulative loss of enzyme activity. The presence of low concentrations of titrating metal inhibitors contacting the reactor over an extended time period can pro-

TABLE 2

HEAVY METAL INHIBITION OF ENZYME ACTIVITY
IN CONTINUOUS PROCESSING REACTORS

Enzyme Reactor Size per Unit Flow Rate (Substrate Residence Time, minutes)	Enzyme Inactivated by Continuous Operation for Time Period Below (grams protein/liter reactor volume)	
	One Month	Three Months
4	22 (30%)	66 (89%)
10	9 (12%)	26 (36%)

20 ppb total irreversibly inhibiting heavy metals, 1 g
enzyme protein deactivated by 10 mg inhibitor, parentheses
give per cent enzyme deactivated based on 75 mg enzyme protein per ml reactor volume.

vide a substantial amount of cumulative enzyme inhibition.
Table 2 illustrates the potential loss of enzyme activity
caused by processing a feed stream containing a total of
20 ppb of heavy metal titrating inhibitors. Let us first
consider an enzyme reactor which has been designed for a
substrate residence time of 4 minutes. In the first month
of continuous operation that enzyme reactor will be contacted by 11,000 reactor volumes of substrate solution.
That amount of substrate solution will contain 220 mg of
inhibitor for each liter of reactor volume. If the titrating inhibitors have been completely reacted in the time
they reside in the reactor and if they bind to the enzyme
with an average value of 10 mg inhibitor per gram of enzyme
inactivated, then 22 grams of enzyme protein will be deactivated per liter of reactor volume. These 22 grams amount
to the loss of 30% of the activity of an immobilized enzyme
containing 75 mg enzyme protein per ml reactor volume.

In the same reactor, after three months of continuous
operation, 66 grams of enzyme protein will be inactivated
per liter of reactor volume. For a reactor containing two
and one half times as much enzyme, a residence time of ten
minutes, the amount of enzyme inactivated will be the same.
However, the fraction inactivated, whether expressed in

grams per liter or percent activity, will be proportionately less. Thus, 9 and 26 grams of enzyme protein will be inactivated after continuous operation for one month or three months, respectively.

The significance of feed purification can perhaps best be exemplified by considering enzyme deactivation caused by a feed stream contaminated with 2mg/liter (2000 ppb) of titrating enzyme inhibitors. After only <u>one day</u> of contamination at this level, the 4 minute residence time reactor would be completely deactivated.

Mercury as a Model Enzyme Inhibitor

Mercury can interact with enzymes in many ways. It can form polyfunctional coordination complexes with several groups on the enzyme or may combine with single sulfhydryl, imidazole, amino or carboxyl groups (8). The interaction may take place at the catalytic site, thus completely inactivating the enzyme. The interaction may also occur at the substrate binding site or a conformational site, producing either partial inactivation (modification of kinetic constants) or complete inactivation. Additionally, mercury can induce changes in enzyme conformation by combining with groups which are essential for maintaining protein structure (24). These conformational changes are generally not reparable by thiol treatment. Mercury may also displace metals in metalloenzymes, thereby deactivating them (11). The classic work of Sumner and Poland (9) showed that the active site of urease contained a sulfhydryl group and that reaction of mercury with this sulfhydryl inactivated the enzyme. Since then mercurial inhibition of many enzymes and the nature of that inhibition has been extensively studied (24).

Four enzymes were selected for mercurial inhibition studies, two commercial grade enzymes: glucose isomerase (Novo SP-103) and lactase (Wallerstein LP-Aspergillus niger) and two research grade enzymes: lactase (Sigma-Saccharomyces fragilis) and alcohol dehydrogenase (Sigma-yeast).

The literature on glucose isomerase inhibition is quite limited. Takasaki et al (16) found the enzyme to be severely inhibited by Hg^{+2}, Ag^{+1} and Cu^{+2}, although the sulfhydryl titrators monoiodoacetate and p-chloromercuribenzoate were not effective inhibitors of this enzyme. These results indicate that mercaptide formation does not play a role in Hg^{+2} inhibition of this enzyme. Mercurial inhibition of yeast alcohol dehydrogenase has been studied extensively and sulfhydryl groups are known to be essential to the activity of this tetrameric metalloenzyme (17). The lactases are of course beta-galactosidases. The sensitivity of beta-galactosidases to heavy metal ions varies greatly, the enzyme from Aspergillus niger exhibits very little heavy metal ion inhibition (18), while the beta-galactosidase from E. coli has been found to be inactivated at low heavy metal concentrations (1 to 10 μM)(19).

Glucose isomerase was extracted from the Novo SP-103 dried cells by incubating one gram of cells in 4 ml of 1M sodium sulfite solution containing 0.1M magnesium sulfate and 0.1M cobalt chloride at 50°C for one hour. The extract was centrifuged to eliminate cell debris and the supernatant fluid used directly for soluble enzyme studies. Immobilized glucose isomerase was prepared by adsorbing the supernatant fluid on 550 Å zirconia clad porous glass which was presilanized to give an alkyl amine derivative. Enzyme assays were performed by incubating the enzyme at 50°C in 0.25M maleate buffer, pH 6.5, containing 10% glucose, 0.1M magnesium sulfate, 0.13M potassium chloride and 2mM cobalt chloride.

Alcohol dehydrogenase was used as purchased. Immobilized alcohol dehydrogenase was prepared by enzyme adsorption on 550 Å zirconia clad porous glass. Enzyme activity was measured spectrophotometrically by determining the rate of NADH production at 25°C in a solution containing 0.55M ethanol, 2mM NAD, 10mM semicarbazide in 0.1M sodium phosphate buffer, pH 8.0. No thiol reagents (DTT, DTE or glutathione) were employed. Immobilized enzyme activity was determined in the recirculation reactor system of Ford et al (12).

The Wallerstein LP Aspergillus niger lactase was used as received. Immobilization of the Aspergillus lactase was achieved by covalently coupling the enzyme to an alkyl amino derivatized 550 Å zirconia clad porous glass with glutaraldehyde. Enzyme activity was assayed at 37°C by incubation in a 10 per cent lactose solution in 0.01M citrate buffer, pH 3.65.

Lactase from Saccharomyces fragilis was used as received. Activity was determined spectrophotometrically by
measuring the rate of p-nitrophenolate ion production. The
enzyme was assayed at 25°C in a 0.01M sodium phosphate buffer, pH 7.0, containing 0.5mM p-nitrophenyl-B-D-galactopyranoside (PNPG) and 10mM magnesium chloride, using the
galactosidase assay procedure of Ford, et al (13). Four
mg of bovine serum albumin were added per mg enzyme to stabilize enzyme activity. It is well to point out that the
use of bovine serum albumin complicates the analysis of
mercury inhibition, since complexation of mercury occurs
with the albumin as well as the enzyme. Hughes (15) noted
that there was only a single sulfhydryl site on the mercap-
talbumin fraction and none on the remaining fractions of
serum albumin. Hughes further noted that this sulfhydryl
is saturated with metal before there is any appreciable
complex formation with other groups on the protein.

Inhibition of Soluble Enzymes by Mercury

Figure 1 illustrates the inhibition of four soluble
enzymes by mercuric chloride. Solutions containing varying
proportions of mercuric chloride and enzyme were incubated
at 25°C for one hour and the enzyme activity subsequently
assayed as described above. It is evident that mercury is
a potent inhibitor of the three enzymes: glucose isomerase,
alcohol dehydrogenase and lactase from Saccharomyces fragilis. In contrast, the lactase isolated from Aspergillus
niger is considerably less inhibited. The amount of mercury required to completely inhibit the Aspergillus enzyme
was found to be 4250 μg mercury per milligram enzyme. For
the three enzymes which were severely inhibited by mercury,
these investigations showed that the loss of enzyme activity was only a function of the ratio of mercury to enzyme,
independent of the absolute mercury concentration and amount
of enzyme. Furthermore, a close examination of the data
reveals that enzyme activity decreases linearly from full
activity to less than 50% activity for all three enzymes.

It can be shown that the fractional enzyme activity
should only be a function of the ratio of inhibitor to enzyme for titrating inhibitors. A titrating inhibitor in-
teracting with strong mercury binding sites essential to
enzyme activity would be expected to initially follow a
linear relation between activity and the ratio of inhibitor

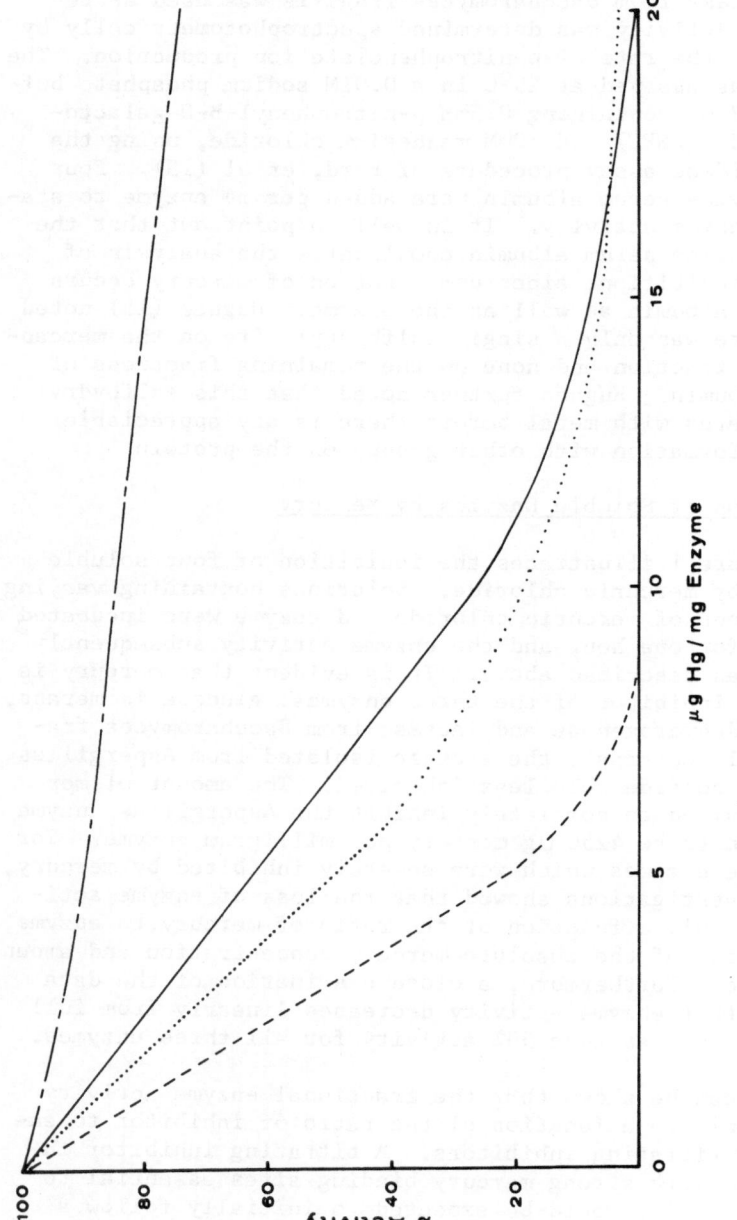

to enzyme. Thus, by extrapolating the linear portion of these experimental curves to intercept the mercury/enzyme ordinate, the intercept gives an apparent titration constant, a stoichiometric mercury/enzyme ratio, operative in this linear region. The titration constants were found to be 6.3, 10.0 and 13.6 μg mercury per mg soluble enzyme for Saccharomyces lactase, alcohol dehydrogenase and glucose isomerase, respectively.

Inhibition of Immobilized Enzymes by Mercury

Each of the four enzymes studied in its soluble form was immobilized and the effect of mercuric chloride on its activity determined. The immobilized enzyme was incubated with mercuric chloride for one hour in a shaker bath at room temperature, washed with buffer and then analyzed for enzyme activity. The effect of mercuric chloride on immobilized glucose isomerase is seen in Figure 2a, while the corresponding effect on the soluble enzyme may be found in Figure 2b.

The effect of mercury on the immobilized enzyme can be seen to be quite different than that on the soluble enzyme. At low inhibitor concentrations, a given quantity of inhibitor ($HgCl_2$) produces a greater inhibition of immobilized enzyme than soluble enzyme. For example, the apparent titration constant, based on initial slope, is 8 μg mercury/mg glucose isomerase for the immobilized enzyme versus 13.6 μg/mg for the soluble enzyme. On the other hand, significant residual enzyme activity (25%) can be found in the immobilized enzyme at mercury/enzyme ratios that would completely inactivate the soluble enzyme.

The extent of inhibition of glucose isomerase by mercury was not altered when the incubation was performed in the assay solution containing 10% glucose, cobalt, magnesium and potassium. The mercury-inhibited enzyme could not be reactivated by simple washing with mercury-free substrate solutions containing Co, Mg and K. Thus, neither high concentrations of substrate nor presence of cobalt or magnesium has any protective effect on the extent of mercury-caused enzyme inactivation. Additionally, preliminary experimental studies with Hg^{203} indicate that mercury taken by the supporting glass matrix is negligible compared with that taken up by the adsorbed enzyme.

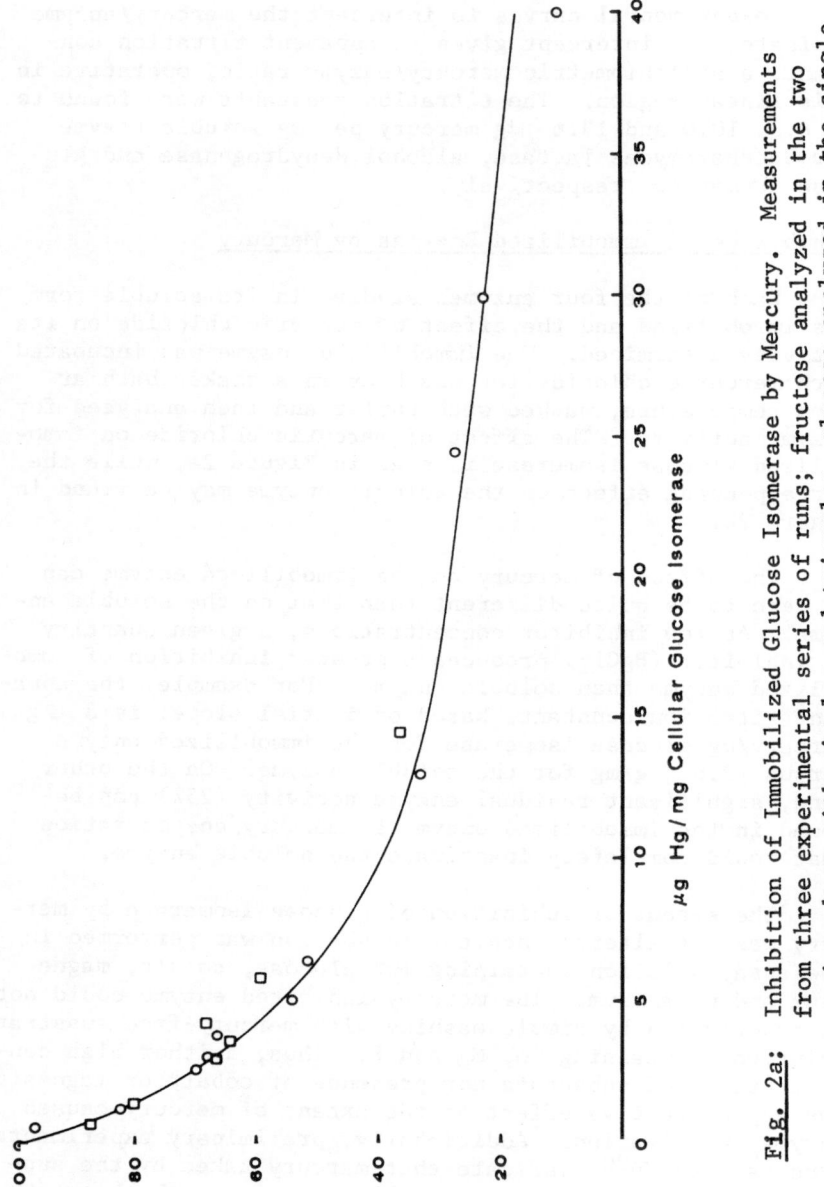

Fig. 2a: Inhibition of Immobilized Glucose Isomerase by Mercury. Measurements from three experimental series of runs; fructose analyzed in the two series shown with circles and triangles, glucose analyzed in the single series shown with squares.

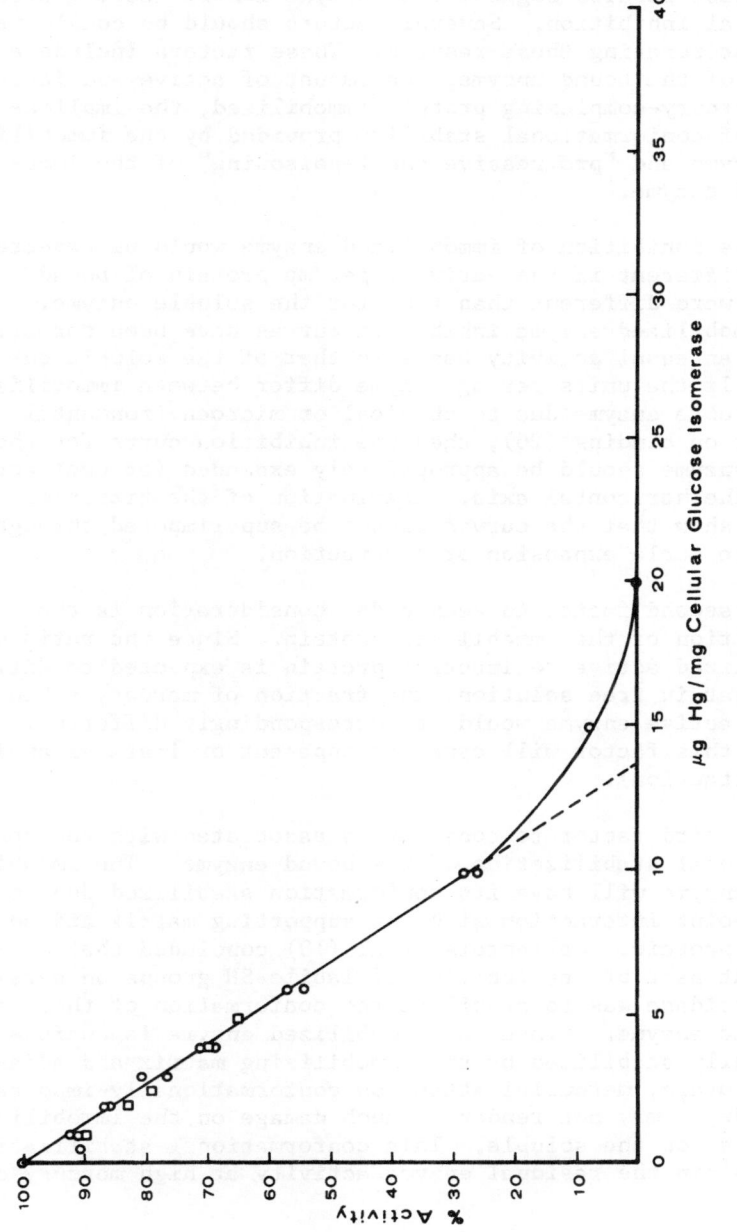

Fig. 2b: Inhibition of Soluble Glucose Isomerase by Mercury, same identification.

These results suggest that enzyme immobilization affects mercurial inhibition. Several factors should be considered when interpreting these results. These factors include activity of the bound enzyme, the amount of active and inactive mercury-complexing protein immobilized, the implications of conformational stability provided by the immobilized enzyme and "progressive shell-poisoning" of the immobilized enzyme.

The inhibition of immobilized enzyme would be expected to be different if the activity per mg protein of bound enzyme were different than that for the soluble enzyme. The immobilized enzyme inhibition curves have been normalized on an equal activity basis to that of the soluble enzyme. If the units per mg enzyme differ between immobilized and soluble enzyme due to chemical or microenvironmental effects on binding (26), then the inhibition curve for the bound enzyme should be appropriately expanded (or contracted) along the horizontal axis. Examination of the titration curves show that the curves cannot be superimposed through ordinate scale expansion or contraction.

A second factor to keep under consideration is the composition of the immobilized protein. Since the ratio of immobilized active to inactive protein is expected to differ from that in free solution, the fraction of mercury taken up by the active enzyme would be correspondingly different. Again, this factor will cause an apparent ordinate expansion (or contaction).

A third factor to consider is associated with the conformational stabilization of the bound enzyme. The immobilized enzyme will have its conformation stabilized due to multi-point interaction with the supporting matrix and adjacent protein. Wallenfels et al (10) concluded that a significant part of the function of labile-SH groups on beta-galactosidase was to stabilize the conformation of this tetrameric enzyme. Since the immobilized enzyme is conformationally stabilized by the immobilizing matrix and adjacent protein, mercurial attack on conformationally-important sulfhydryls may not render as much damage on the immobilized enzyme as on the soluble. This conformational stabilization may explain the residual enzyme activity at high mercury content .

One effect that is known to cause concave-upward activity versus titrating inhibitor curves in heterogenous catalysis is that of "progressive shell-poisoning" (20). In a progressive shell poisoning situation, the mercuric chloride would preferentially inhibit enzyme near the outside of the immobilized enzyme particles first. As more mercury accumulates in the immobilized enzyme, this layer of inhibited enzyme will grow progressively larger, forming an inwardly-growing "shell". Thus, the active enzyme in the core will be encased in a slowly growing annular region of inactive enzyme. Concave upward curvature is produced if the outer inactive enzyme layer provides sufficient substrate diffusional resistance, thus causing the enzyme activity to drop more rapidly with mercury content than if the mercury were only acting to titrate the enzyme.

A New Selective Heavy Metal Adsorbent

Recognizing the problem of enzyme inactivation by heavy metals, an examination of several currently available approaches to its solution was made. Heavy metals can be removed by ion exchange columns or by selective precipitation. Protective agents that form tight complexes with the metal may be used if these metal-agent complexes have higher binding constants than the corresponding metal-enzyme inhibition complexes.

Each of these approaches has significant drawbacks. The heavy metal selectivity of ion exchangers leaves much to be desired. Heavy metal selectivity is especially important, as heavy metals may be in the part per billion range while other ionic species are at millimolar or higher concentrations. Selective precipitation and subsequent filtration of part per billion quantities of an inhibitor is difficult, further compounded by the spectrum of heavy metals anticipated. Selective precipitation is costly and carries an additional debit of introducing a precipitating agent. Protective agents, like EDTA, dithiothreitol and cysteine, when added continuously, represent an expensive proposition with respect to both the cost of the agent and the cost of separating the agent from the reaction mixture.

The development of a new class of heavy metal selective absorbents was therefore initiated. These adsorbent are prepared by covalently attaching sulfhydryl groups to in-

organic matrices via silane intermediates. The resultant
heavy metal adsorbent tightly binds mercury by a mercaptide
complex that is essentially irreversible. The capacity of
this heavy metal adsorbent is dependent on the sulfhydryl
content. The higher sulfhydryl content, silica-alumina
based adsorbent designated as ASA-100 has a capacity of
106 mg Hg/g adsorbent, while the lower sulfhydryl content
diatomite-based AD-30 has a capacity of 12 mg Hg/g.

These new heavy metal adsorbents were found to be quite
stable in mild oxidizing environments. Their behavior con-
trasts to that of thiol-based redox polymers which have also
been used for selective heavy metal binding (21). The adsorbent AD-30 was found to be resistive to oxidation by
ferricyanide, dilute hypochlorite or dissolved oxygen.
However, incubation of the adsorbent in strong oxidizing
agents, such as acid permanganate or perchloric acid, substantially decreased the capacity for subsequent heavy metal
adsorption. On the other hand, incubation in reducing
agents, e.g. sodium borohydride, stannous chloride, sodium
dithionite or sodium bisulfite, did not affect subsequent
heavy metal adsorption.

One important characteristic of an adsorbent to be
used for purification of immobilized enzyme reactor feed
solutions in heavy metal selectivity. It was found that
large excesses of common metal ions do not significantly
reduce the capacity of the adsorbent for mercury. Table
3 gives the relative performance of heavy metal adsorbent
AD-30 for the adsorption of mercury from aqueous solutions
containing base metal ions. The initial solution in all
cases contained 100 ppb of $HgCl_2$.

Adsorbent Kinetics and Adsorber Column Behavior

The kinetics of mercuric chloride adsorption was found
to be first order in mercury. For small adsorbent particles, e.g. 20 microns AD-30, the rate was also found to be
linearly dependent on the amount of mercury capacity remaining. For larger particles the rate of adsorption is
dependent in a nonlinear fashion on both the amount of
mercury capacity remaining and the size of the adsorbent
particle. These experimental data suggested that the description of adsorbent behavior by means of a "progressive
shell" or "pore-mouth poisoning" model might be useful.

TABLE 3

HEAVY METAL SELECTIVITY

Added Metal Ion	Relative Amount of Hg Removed
None	1.00
0.01 M HaCl	0.95
0.01 M CaCl$_2$	0.95
0.01 M MgCl$_2$	0.98
0.01 M MnSO$_4$ + 0.01 M NaCl	0.94

Consequently, a progressive shell mathematical model for adsorbent behavior was developed. In that model, mercuric chloride initially reacts with active groups on the periphery of the porous adsorbent, forming an outer layer or shell of adsorbed mercury. Subsequent mercuric chloride must diffuse through this fully adsorbed outer layer into the active core of the adsorbent. As more mercuric chloride arrives, the outer layer grows until all the active groups on the particle have been reacted with mercury.

The mathematical model was used to predict F, the ratio of the rate of adsorption for adsorbent containing an amount of mercury θ to that for fresh adsorbent. Theta, θ, is the fraction of the maximum mercury binding capacity adsorbed. Experimental verification of this model may be seen in Figure 3. The solid lines shown in this figure represent a priori model prediction for the particle sizes employed (23, 97 and 326 microns). The agreement between progressive shell model predictions and experimental results is quite favorable.

The behavior of a column of heavy metal adsorbent was described by a digital computer simulation using the progressive shell model discussed above. The column performance may be described by the following partial differential equations:

$$u \frac{\partial C}{\partial z} + \varepsilon \frac{\partial C}{\partial t} = \rho_b \frac{\partial Q}{\partial t} \tag{1}$$

$$\frac{\partial Q}{\partial t} = \tau_o \eta_o FkC \tag{2}$$

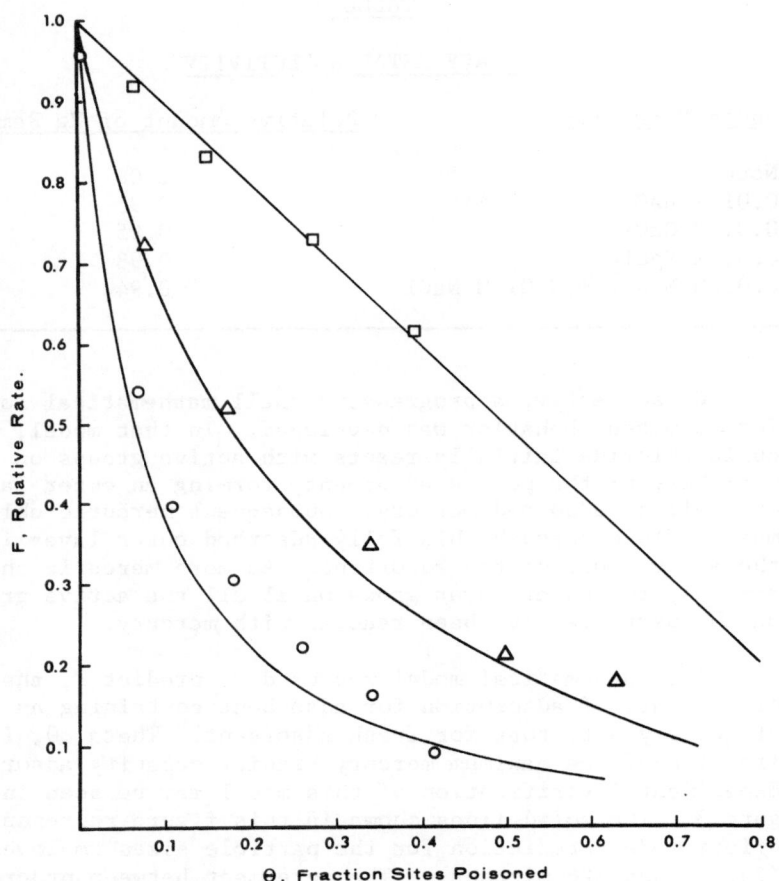

<u>**Fig. 3:**</u>   Effect of Progressive Shell Poisoning on the Rate
            of Adsorption. Circles 326 μm, triangles 97 μm,
            and squares 23 μm diameter adsorbent AD-30

where Q represents the metal adsorbed per unit mass of adsorbent, C the metal concentration, ρ_b the adsorbent bulk
density, ϵ the external void fraction in the column, η_o the
internal effectiveness factor for fresh adsorbent, τ_o the
external effectiveness factor, and $\tau_o \eta_o FkC$ the rate of adsorption. The boundary conditions are such that at z = 0,
C = C_i for all , and at =0, C and Q are 0 for z greater
than 0. A numerical algorithm (22) for the solution of the

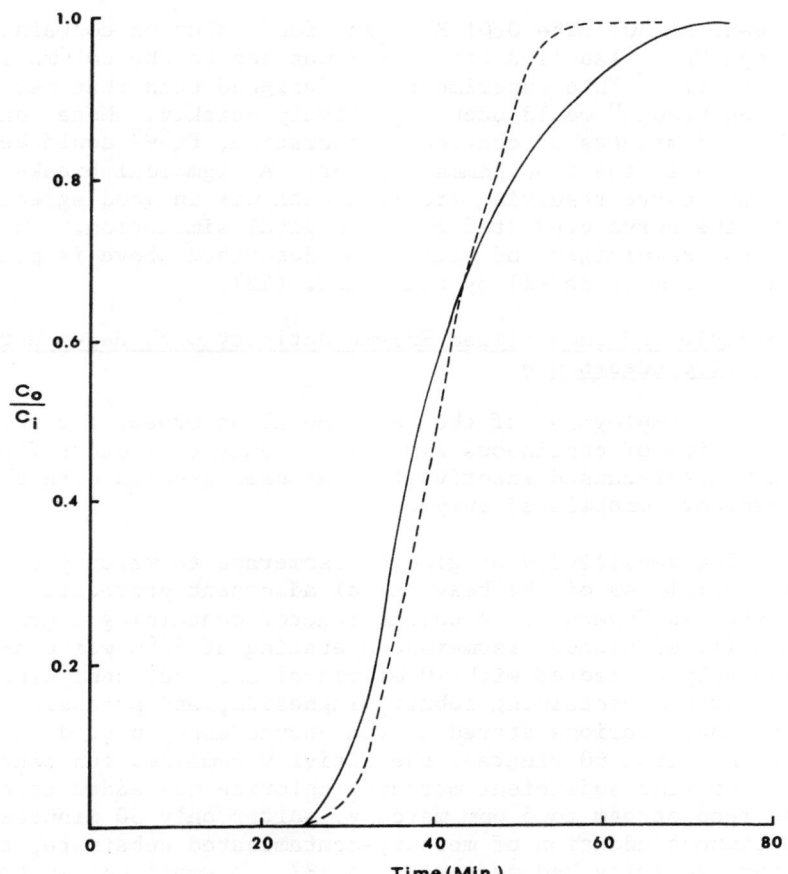

$$\frac{C_o}{C_i}$$

Time (Min.)

Fig. 4: Adsorber Column Behavior. Mercuric chloride breakthrough curve on 97 m diameter AD-30.

two coupled hyperbolic partial differential equations was
formulated to predict adsorbent column behavior.

Experimental verification of column performance was
conducted with several adsorbent columns. Figure 4 shows
the computer simulation (dashed line) and the experimental
results (solid line) obtained from a 13 cm long by 0.2 cm
diameter column containing 184 mg of 97 micron diameter ad-

sorbent AD-30. The 0.01 M saline feed solution containing
10 ppm $HgCl_2$ labelled with Hg^{203} was fed to the column at
3.7 ml/min. This experiment was designed such that mercury
"breakthrough" would occur relatively quickly. Note that
after 25 minutes of continuous operation, Hg^{203} could be
detected in the flow gamma counter. A sigmoidal breakthrough curve resulting from the data was in good agreement
with the curve predicted by the digital simulation. The
digital simulation and math model described above is presented in more detail by Swan et.al. (23).

Protection of Immobilized Enzyme Activity with Heavy Metal Selective Adsorbents

The employment of the heavy metal adsorbent for the
protection of continuous immobilized enzyme reactors from
heavy metal-caused inactivation has been studied with three
different immobilized enzymes.

The sensitivity of glucose isomerase to mercury and
the usefulness of the heavy metal adsorbent protection can
be seen in Figure 5. A column reactor containing 1 gram of
immobilized glucose isomerase operating at $50^{o}C$ was continuously contacted with 10 ml/min of a 10 per cent glucose
feed stream containing cobalt, magnesium, and potassium at
the concentrations stated in the enzyme assay procedure.
For the first 60 minutes, the activity remained constant.
At that time sufficient mercuric chloride was added to bring
the feed stream to 5 ppm mercury. After only 30 minutes of
continuous addition of mercury-contaminated substrate, the
enzyme activity had decreased to 48%. A small column containing 2 g of heavy metal adsorbent AD-30 was then added
just upstream of the enzyme reactor. As can be seen from
Figure 5, the adsorbent removed mercuric chloride from the
feed stream, thus preventing heavy metal-caused activity
loss. The activity loss from 90 minutes to 240 minutes was
only one per cent, which can be attributed to slow enzyme
denaturation at these conditions. At 240 minutes, the adsorbent column was removed from the system and the mercury-
contaminated feed stream, which had been flowing continuously through the adsorbent column was allowed to again
directly contact the enzyme reactor. A rapid inactivation
subsequently occurred with the enzyme reactor activity de-
clining to 3 per cent of its initial value at the end of
the 6 hour experiment.

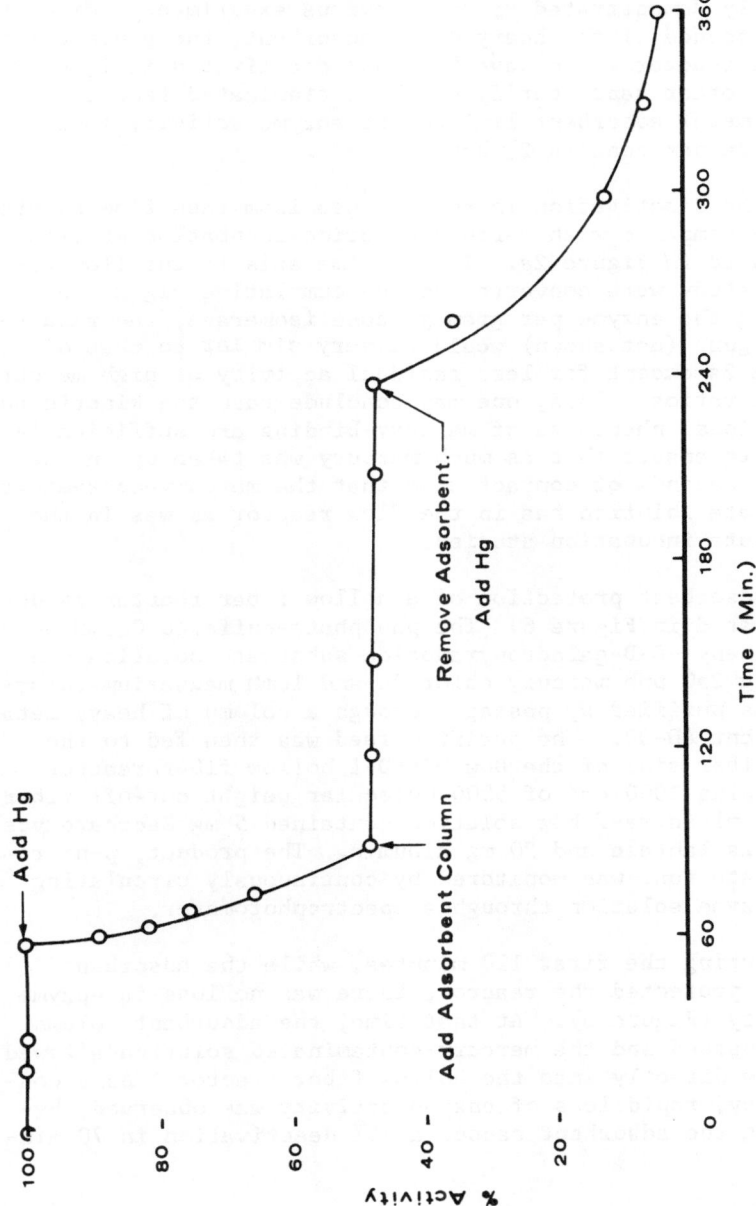

Fig. 5: Protection of Immobilized Glucose Isomerase by Heavy Metal Adsorbent AD-30.

The necessity for eliminating deleterious heavy metals is amply demonstrated by the previous experiment. Without the presence of the heavy metal adsorbent, the glucose isomerase reactor would have been 90% deactivated in 1½ hours. On the other hand, purifying the contaminated feed with the heavy metal adsorbent limited the enzyme activity loss to only one per cent in 2½ hours.

The deactivation in the glucose isomerase flow reactor may be compared with mercury chloride incubation studies presented in Figure 2a. If the time axis in the flow reactor study were converted to the cumulative μg Hg contacting the enzyme per gram glucose isomerase, the resulting figure (not shown) would be very similar to that of Figure 2a except for less residual activity at high mercury/enzyme ratios. Thus, one may conclude that the kinetic and diffusional phenomena of mercury-binding are sufficiently rapid to ensure that as much mercury was taken up in the twelve seconds of contact time that the mercury-contaminated substrate solution has in the flow reactor as was in the 60 minute incubation studies.

Adsorbent protection of a hollow fiber reactor is demonstrated in Figure 6. The phosphate-buffered 0.5mM p-nitrophenyl-B-D-galactopyranoside substrate solution containing 250 ppb mercury chloride and 10mM magnesium chloride was purified by passage through a column of heavy metal adsorbent AD-30. The purified feed was then fed to the extrafiber side of the Dow b/HFD-1 hollow fiber reactor containing 1000 cm^2 of 5000 molecular weight cut-off fibers. The 20 ml intra-fiber solution contained 5 mg Saccharomyces fragilis lactase and 20 mg albumin. The product, p-nitrophenolate ion, was monitored by continuously circulating the enzyme solution through a spectrophotometer.

During the first 110 minutes, while the adsorbent column protected the reactor, there was no loss in enzyme activity (Figure 6). At that time, the adsorbent column was bypassed and the mercury-contaminated solution allowed to flow directly into the hollow fiber reactor. As a consequency, rapid loss of enzyme activity was observed; bypassing the adsorbent caused a 45% deactivation in 70 minutes.

The effect of heavy metals which are self-reversing inhibitors on enzyme reactor performance is quite different

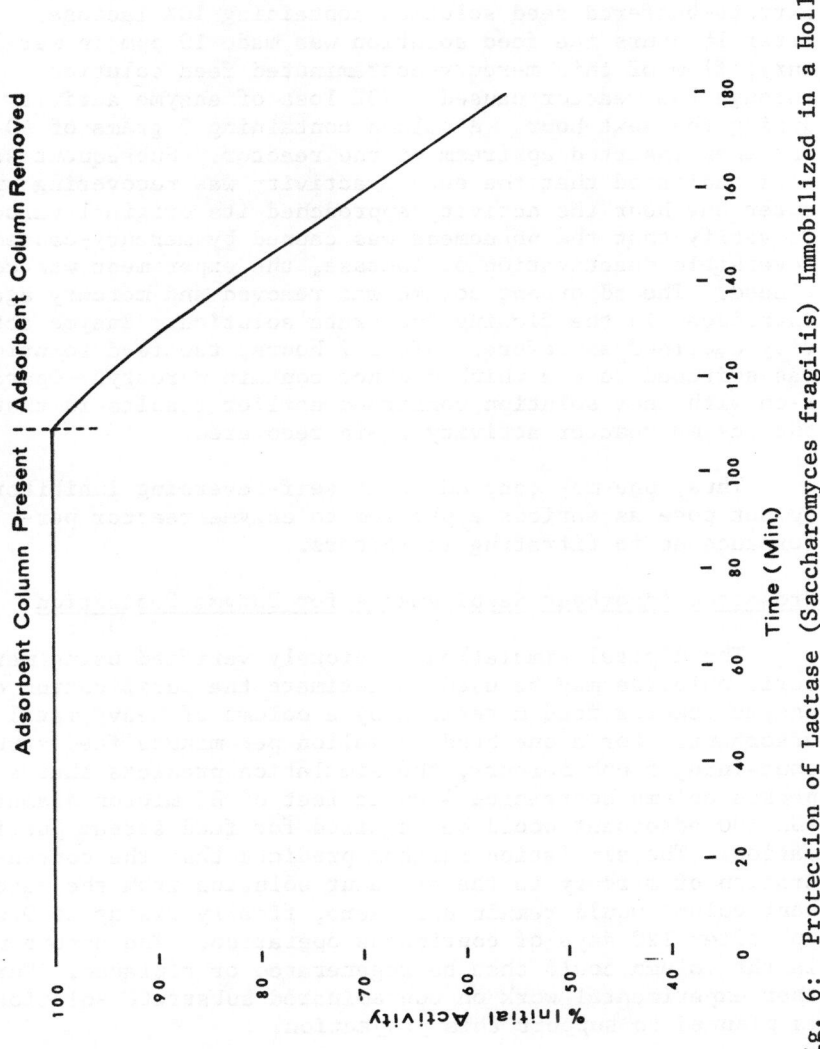

Fig. 6: Protection of Lactase (Saccharomyces fragilis) Immobilized in a Hollow Fiber Reactor by Heavy Metal Adsorbent AD-30.

from the titrating inhibitor situations just presented. An
illustration of this effect may be seen in Figure 7. A
small reactor containing 160 mg of glass-bound Aspergillus
niger lactase was operated at steady state with pH 3.7
citrate-buffered feed solution containing 10% lactose.
After 1½ hours the feed solution was made 10 ppm in mercury; flow of this mercury-contaminated feed solution
through the reactor caused a 40% loss of enzyme activity
during the next hour. A column containing 2 grams of AD-30
was then inserted upstream of the reactor. Subsequent samples indicated that the enzyme activity was recovering and
after one hour the activity approached its original value.
To verify that the phenomena was caused by mercury-caused
reversible deactivation of lactase, the experiment was continued. The adsorbent column was removed and mercury again
introduced in the flowing substrate solution. Enzyme activity declined as before. After 7 hours, the feed solution
was switched to one which did not contain mercury. Operation with that solution confirmed earlier results in that
the enzyme reactor activity again recovered.

Thus, one may conclude that self-reversing inhibitors
do not pose as serious a problem to enzyme reactor performance as to titrating inhibitors.

Projected Adsorbent Requirements for Enzyme Protection

The digital simulation previously verified using mercuric chloride may be used to estimate the purification of
enzyme reactor feed materials by a column of heavy metal
adsorbent. For a one hundred gallon per minute feed stream
containing 5 ppb mercury, the simulation predicts that a
packed column containing 4 cubic feet of 80 micron diameter
ASA 100 adsorbent would be required for feed stream purification. The simulation further predicts that the concentration of mercury in the effluent solution from the adsorbent column would remain near zero, finally rising to 0.05
ppb after 120 days of continuous operation. The adsorbent
in the column could then be regenerated or replaced. Further experimental work on contaminated substrate solutions
is planned to support this projection.

ACKNOWLEDGEMENT

The authors would like to thank the National Science
Foundation Rann Program (Grant GI 34974) for support of

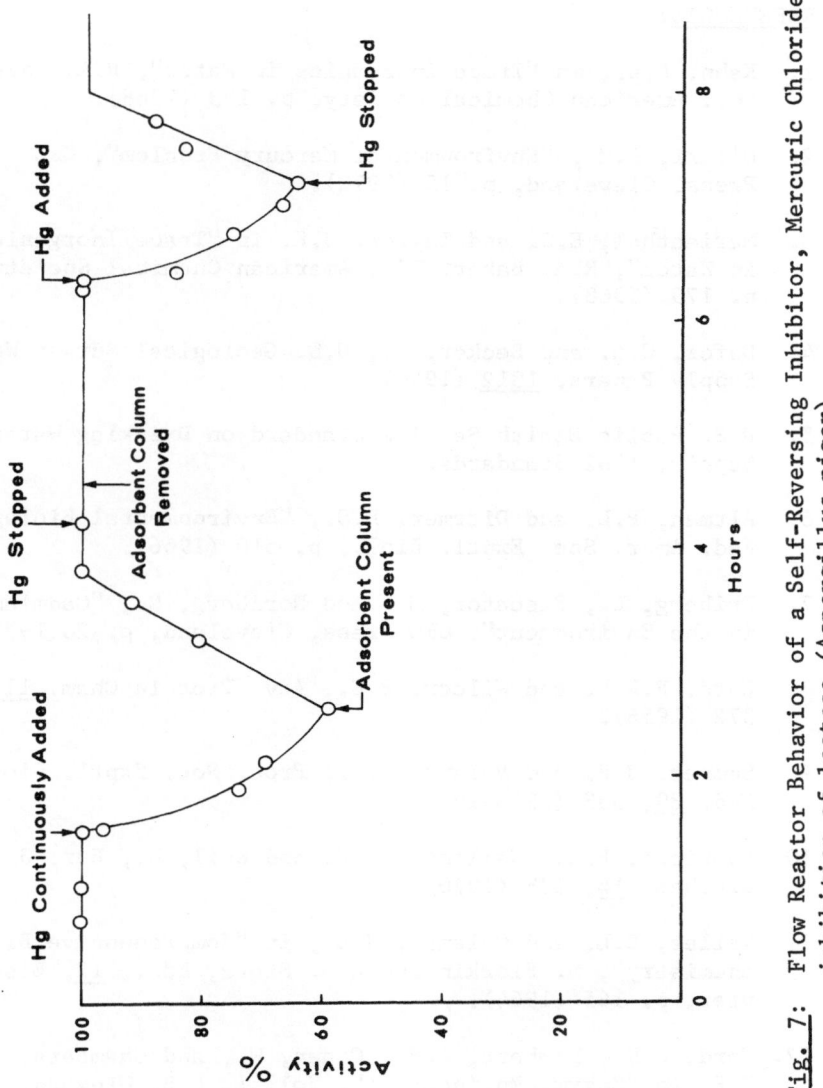

Fig. 7: Flow Reactor Behavior of a Self-Reversing Inhibitor, Mercuric Chloride inhibition of lactase (Aspergillus niger).

this research and the NSF Office of International Programs for supporting its presentation.

REFERENCES:

1. Kahn, H.L., in "Trace Inorganics in Water", R.A. Baker, Ed., American Chemical Society, p. 193 (1968).

2. D'Itri, F.M., "Environmental Mercury Problem", CRC Press, Cleveland, p. 15 (1972).

3. Marienthal, E.J. and Taylor, J.K. in "Trace Inorganics in Water", R.A. Baker, Ed., American Chemical Society, p. 179 (1968).

4. Dufor, C.N. and Becker, E., U.S. Geological Survey Water Supply Papers, 1812 (1964).

5. U.S. Public Health Service Standard on Drinking Water Supply, 1962 Standards.

6. Altman, P.L. and Dittmer, D.S., "Environmental Biology", Fed. Amer. Soc. Exptl. Biol., p. 510 (1966).

7. Friberg, L., Piscator, M., and Nordberg, G., "Cadmium in the Environment", CRC Press, Cleveland, p. 26(1971).

8. Gurd, F.R.N. and Wilcox, P.E., Adv. Protein Chem. 11, 372 (1956).

9. Sumner, J.B. and Poland, L.O., Proc. Soc. Exptl. Biol. Med. 30, 553 (1923-1933).

10. Lootiens, F.G., Wallenfels, K. and Weil, R., Eur. J. Biochem. 14, 138 (1970).

11. Vallee, B.L. and Coleman, J.E., in "Comprehensive Biochemistry", M. Florkin and E.H. Stotz, Eds., 12, Elsevier, p. 165 (1964).

12. Ford, J.R., Lambert, A.H., Cohen, W., and Chambers, R.P. in "Enzyme Engineering", Vol. 1, L.B. Wingard, Ed., Wiley, p. 267 (1972).

13. Ford, J.R., Nunley, J.A., Li, Y.T., Chambers, R.P. and Cohen, W., Anal. Biochem. 54, 120 (1973).

14. Ford, J.R., Ph.D. Dissertation, Tulane University, New Orleans, Louisiana (1972).

15. Hughes, W.L., Cold Spring Harbor Symp. Quan. Biol. 14, 79 (1950).

16. Takasaki, Y., Kosugi, Y. and Kanbayashi, A., in "Fermentation Advances", D. Perlman, Ed., Academic Press, p. 561 (1969).

17. Sund, H. and Theorell, H., The Enzymes 7, 25 (1963).

18. Wallerstein Corporation, personal communication.

19. Wallenfels, K. and Malhotra, O.P., The Enzymes 4, 409 (1960).

20. Wheeler, A., Adv. Catalysis 3, 249 (1951).

21. Kun, K.A., U.S. Patent 3,278,487 (October 11, 1966).

22. von Rosenberg, D.U., Chambers, R.P., Swan, G.A. and Parnell, M.B., "Numerical Solution of Fixed Bed Adsorption", paper 7a, Salt Lake City AIChE (1974).

23. Swan, G.A., McElrath, K.O., Baricos, W.H.,Cohen, W. and Chambers, R.P., "Heavy Metal Removal from Aqueous Media", accepted for publication in AIChE Symposium Series, T. Vermeulen, Ed. (1975).

24. Webb, J.L., "Enzyme and Metabolic Inhibitors", Volume II, p. 729, Academic Press (1966).

25. Cranston, R.E. and Buckley, D.E., Envir. Sci. Tech. 6, 274 (1972).

26. Goldman, R., Goldstein, L. and Katchalski, E., in "Biochemical Aspects of Reactions on Solid Supports", G.R. Stark, Ed., p. 1, Academic Press (1971).

EXTERNAL DIFFUSION OF ENZYMES ENTRAPPED WITH CALCIUM

ARGINATE IN PACKED BEDS

Susumu Fukushima

Department of Chemical Engineering, Kansai University, Suita, Osaka 564, Japan

Immobilized enzymes have actively been investigated as heterogeneous catalysts for industrial applications. The packed bed is one of the typical reactor for these. In practical operation of immobilized enzymes as well as ion exchanges the liquid flow rate is very low and the particle diameter is generally much smaller than common industrial catalysts. In a region around the exterior of the particle where the flow rate is very low, the substrate must be transported to the particle where the catalyst effects occur. Since this transfer rate of substrate may be decreased significantly by the external diffusion. The effect of external diffusion rate of substrate is thus required for purpose of plant design and operation in the packed beds of immobilized enzymes.

Review for the empirical equations of the mass transfer coefficient on the external diffusion in gas or liquid flow in packed beds has presented by Karabelas et al (5) and Coloquhoun-Lee and Stepanck (2). Some data on liquids in the low Reynolds numbers are available in dissolution of 2-naphthol pellets(3,7), benzoic acid pellets(3) or spheres (11,15,16), and succinic and salicyclic acid pellets(4). Pfeffer(9) presents the theoretical equation,

$$Sh = 1.26 \frac{(1-\epsilon)^{5/3}}{2 - 3(1-\epsilon)^{1/3} + 3(1-\epsilon)^{5/3} - 2(1-\epsilon)^2}^{1/3} \, Sc^{1/3} \, Re^{1/3} \tag{1}$$

which is in agreement with the dissolution data in literature. Karabelas et al (5) give the equation

$$Sh = 4.58Re^{1/3}Sc^{1/3} \tag{2}$$

in the large Pe from the experiment with electrochemical techniques.

Very little information is available in immobilized enzymes. Kittrell and colleagues (10,12) have recently reported that the mass transfer coefficient of substrate is proportional to a half power of the superficial mass velocity of liquid in $0.05 \sim 1mm$ spheres of glucose oxidase immobilized on porous glass and catalase immobilized on the nickel-silica alumina in the low Re ranging from 0.2 to 119.

Obata et al (8) report on immobilized method that the enzyme is adsorbed on a carrier and then entrapped by calcium arginate. The arginate enzyme is innoxious for human body. It is very different from common industrial catalysts, for the particle of arginate enzyme is an elastic body allowable to be deformed in mounting operation of packed beds.

The present work was attempted to obtain the empirical equation for mass transfer coefficient in the external diffusion around the immobilized glucose oxidase spheres adsorbed on $60 \sim 120$ mesh γ-alumina and then entrapped with calcium arginate, and the immobilized invertase entrapped with calcium arginate in the low Re. Various shapes of immobilized enzymes are mounted in packed beds for industrial application. The shape factor is thus of importance for external diffusion. It was also investigated on the glucose oxidase immobilized on 1/8"γ-alumina cylinder.

Oxidation of glucose with soluble and immobilized glucose oxidase in gas-liquid contactors are also interesting for reactor design. In order to investigate the reaction rate equation, this oxidation was also studied in the stirred tank having the free gas-liquid interface.

MATHEMATICAL MODEL

(1) <u>Oxygen Absorption into Glucose Solution with Soluble Glucose Oxidase</u>. The rate of chemical reaction in high glucose concentration is obtained by

$$R_A = kC_{Eo}C_A \tag{3}$$

When the chemical reaction rate is slow, molecular oxygen diffuses in the liquid through the gas-liquid interface without reaction and reacts with glucose in the bulk of liquid. For ideal mixing in the continuous gas-liquid stirred tank, the material balance of oxygen is expressed at the steady state as follows:

$$FC_{Ai}(A_o - A_{in}) = k_1^* a C_{Ai}(1 - A_o)V_1 - R_A V_1 \qquad (4)$$

$$\text{or } k_1^* a (1 - A_o) - (A_o - A_{in})/\tau = k C_{Eo} A_o \qquad (5)$$

In a batch system, $\tau \to \infty$

$$k_1^* a (1 - A_o) = k C_{Eo} A_o \qquad (5')$$

(2) <u>Oxygen Transfer in Suspension of Immobilized Glucose Oxidase</u>. The material balance of oxygen is given in the continuous gas-liquid stirred tank suspended the immobilized glucose oxidase for ideal mixing at the steady state as follows:

$$k_1^* a (1 - A_o) - (A_o - A_{in})/\tau = K a_s A_o \qquad (6)$$

where

$$a_s = \frac{\sum_n \pi d_p^2}{V_1} = \frac{n \pi d_p^{-2}}{V_1} \qquad (7)$$

and K is the apparent specific rate of particles immobilized enzymes.

(3) <u>Packed Bed</u>. At the steady state, the external diffusion rate of substrate A around the particles of immobilized enzymes is equal to the overall reaction rate in the particles which is the apparent Michaelis-Menten type without inhibition as follows:

$$N_A = k_{1s}^* a_s (C_A - C_{As})V_1 = \frac{k' C_{Eo} C_{As}}{K_m' + C_{As}} V_1 \qquad (8)$$

Thus

$$(C_{Ao} - C_{As}) = \frac{k' C_{Eo}}{k_{1s}^* a_s} \frac{C_{As}}{K_m' + C_{As}} = \frac{\alpha C_{As}}{K_m' + C_{As}} \qquad (9)$$

where

$$\alpha = \frac{k'C_{Eo}}{k^*_{1s}a_s} \tag{10}$$

The substrate concentration at the particle surface C_{As} is given from Eq. 9.

$$C_{As} = 1/2 \ (C_A - K'_m - \alpha) + . \ \overline{(C_a - K'_m - \alpha)^2 + 4K'_m C_A} \tag{11}$$

In a packed bed as a plug flow, the material balance of substrate A can therefore be expressed by

$$-\frac{dC_A}{dV_1} = \frac{k^*_{1s}a_s}{F} \ [C_A - 1/2\{(C_A - K'_m - \alpha) + \sqrt{(C_A - K'_m - \alpha)^2 + }$$

$$\overline{4K'_m C_A}\}] \tag{12}$$

If the value of α is experimentally equal to zero for the immobilized particles suspended in a vigorous stirred tank, the apparent specific rate k' and the apparent Michaelis constant K'_m are respectively determined. The value of k^*_{1s} in a packed bed may be obtained from integration of Eq. 12 by the analog computer with boundary condition, $C_A = C_{Ain}$ for $z = 0$.

When $K''_m \gg C_{As}$,

$$N_A = Ka_s C_A V_1 \tag{13}$$

where

$$1/K = 1/k^*_{1s} + K'_m a_s / k' C_{Eo} = 1/k^*_{1s} + 1/k'' \tag{14}$$

and

$$k'' = \frac{k'}{K'_m} \frac{C_{Eo}}{a_s} \tag{15}$$

The material balance of A in a plug flow is simply expressed as

$$C_{Ao}/C_{Ain} = \exp(-K\frac{\varepsilon V}{F}) = e^{-K\tau} \tag{16}$$

EXPERIMENTAL PROCEDURE

Preparation of Immobilized Enzymes

(1) <u>Arginate Glucose Oxidase</u>. Glucose-catalase (2:1)
(Dawe's Lab., Technical E) was solved in buffer solution
and adsorbed on 0.25 g of 60~120 mesh γ-Al$_2$O$_3$ (Nikki Kagaku Co., surface 200 m^2/g, void fraction 60%). The slurry,
which the alumina was suspended in 25g of 1.5wt% sodium arginate, was dropped through a nozzle into 250g of 17wt% calcium chloride solution under magnetic stirring and the drops
were gradually solidified as the immobilized glucose oxidase entrapped with calcium chloride. The spheres of immobilized glucose oxidase were washed and stored in 0.02M
glucose-buffer solution at 4°C. The diameters of spheres
were 1.3 and 3mm.

(2) <u>Arginate Invertase</u>. The solid of crude invertase
(Sigma Co., Grade V) was entrapped with calcium chloride
instead of the alumina adsorbed the soluble enzymes mentioned above. The diameters of arginate invertase spheres
were 0.4 and 3mm.

(3) <u>Alumina Glucose Oxidase</u>. Glucose oxidase-catalase
was dissolved in buffer solution and adsorbed on 1/8" γ-
Al$_2$O$_3$ cylinder for one day at 4°C. To remove free glucose
oxidase. The cylinders of immobilized glucose oxidase were
mounted in baskets and washed in the first stage continuous
stirred tank as analous to the Carberry's reactor (1). The
content of glucose oxidase was roughly 5%.

Reactor System

<u>Stirred Tank</u>. The glass stirred tank with jacket was
T = 8.3cm in diameter with 4 buffles, 0.1T in width. The
agitator was 6 blades turbine impeller, 0.5T in diameter,
mounted at T/2 in height from the bottom of tank. The
liquid height was equal to T. The stirred tank had the
free gas-liquid interface for oxygen absorption with soluble glucose oxidase and pure oxygen transfer in suspension
of arginate enzyme. The solution was mainly 0.2M glucose
included glucose oxidase for absorption. In a continuous
system, the liquid stored in the head tank under nitrogen
gas was supplied to the stirred tank through the rotameter
and pure oxygen gas was also supplied to the upper part of
the tank from the gas holder through the soap flow meter.

The oxygen absorption rate in the tank was directly measured
by the soap flow meter and determined from the difference
of oxygen concentration in the bulk of liquid between inlet
and outlet or the stirred tank by the oxygen analyzer.

In a batch system, the oxygen absorption rate was determined from the oxygen concentration in the bulk of liquid at the pseudo steady state. Both systems were maintained at $25^{\circ}\pm0.2^{\circ}C$. The physical absorption of oxygen
was also investigated in 0.2M glucose solution included
glucose oxidase at pH 1.8, because the enzyme was inactive.
The absorption rate was measured in suspension of arginate
spheres without glucose oxidase. The effect of glucose and
ion on the solubility of oxygen was estimated by van Kreve-
len-Hoftijzer method (14).

The oxygen transfer rate for alumina glucose oxidase
was also determined in a continuous liquid-solid stirred
reactor as analogous to the Carberry's reactor for gas-
solid catalyst reaction. The four baskets, 0.4T in width
were mounted at the rotating shaft between 4 blades turbine
impellers, 0.6T in diameter. The cylinders of alumina glucose oxidase were arranged in the baskets. The rate was
determined from the difference of the oxygen concentration
between inlet and outlet.

At the unsteady state the sucrose transfer rate in
suspension of crude and arginate invertase was determined
from the reduced sugar by Somogi's method.

Packed Beds. The beds were 1.2, 1.9 and 2.7cm in diameter. The same sized arginate spheres or γ-Al$_2$O$_3$ cylinders without enzymes as immobilized enzymes were mounted in
the upper and the lower parts of beds. The liquids are
0.3M glucose solution saturated with air or 0.06M sucrose
solution contained the buffer solution. The Schmidt number
is 4.9×10^2 for glucose solution and 1.87×10^3 for sucrose
solution. The liquid flow was directed upward in the packed
bed through the glass filter and rotameter.

RESULTS AND DISCUSSION

Stirred tank. All of these continuous stirred tanks
were identified to be ideal mixing by virtue of the pulse
response techniques in these experimental conditions. In

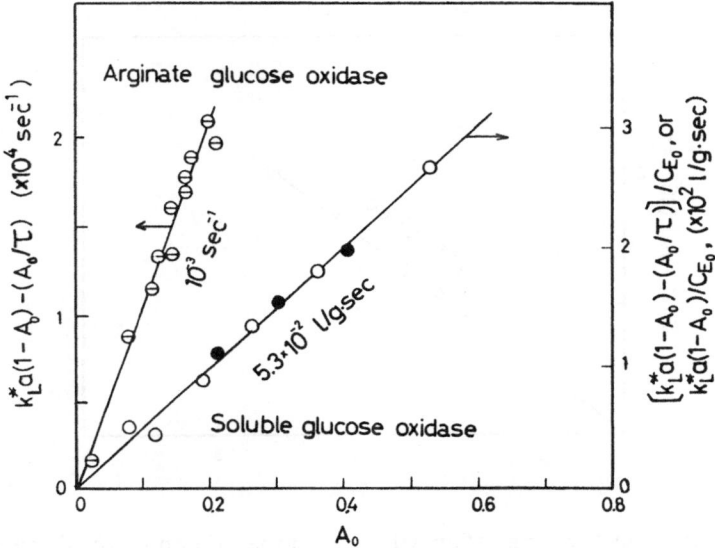

<u>Fig. 1</u>: Oxygen absorption with soluble glucose oxidase and
        oxygen transfer in suspension of arginate glucose
        oxidase (25°C)
        ● continuous stirred tank, $C_{Eo} = 1.3 \times 10^{-2}$ g/1,
        $C_G = 0.2M$, $N = 241 \sim 620$ r.p.m., $\tau = 594 \sim 1200$ sec.,
        pH = 3.8,
        ○ batch stirred tank $C_{Eo} = 1.3 \times 10^{-2} \sim 1.3 \times 10^{-1}$ g/1,
        $C_G = 0.2M$, $N = 150 \sim 620$ r.p.m., pH = 3.8~7,
        ⊖ continuous stirred tank, dp = 0.29 cm, $a_S = 0.44$
        cm^{-1}, n = 1063 1^{-1}, E = $2.2 \times 10^{-4} \sim 5.6 \times 10^{-4}$ g/n,
        $C_G = 0.2 \sim 0.5M$, N = 241,371 r.p.m, $\tau = 587 \sim 930$ sec.,
        pH = 6.

oxygen absorption, the enhancement factors were actually
equal to 1 in the stirred speeds ranging from 150 to 640
r.p.m. where the concentration of glucose oxidase was below
0.13g/1. Lee and Tsao (6) reported that the enhancement
factor in a batch stirred tank was higher than the value estimated by chemical absorption theory. This is because the
volumetric coefficient was taken the value at $C_{Eo}=0$ in the
extrapolation of the plots of the coefficient with chemical
reaction against the concentration of soluble glucose oxidase.

    In order to investigate the enhancement factor, it is
important that the physical absorption must be measured in
the presence of the same amounts of glucose oxidase as chemi-

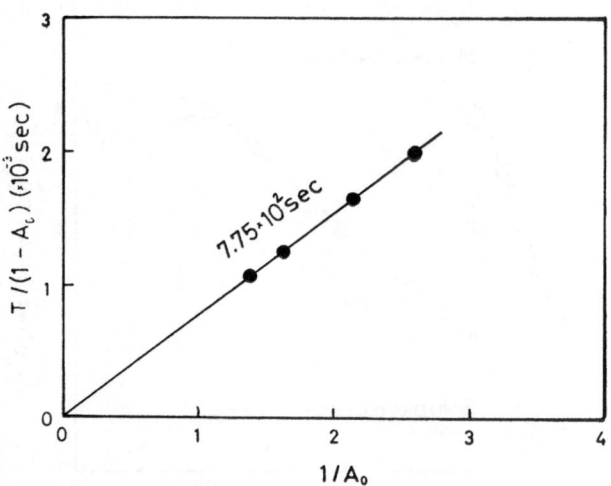

Fig. 2: Oxygen transfer with alumina glucose oxidase cylinders in a basket-type continuous stirred tank
(25°C).
$dp = 1/8"$, $a_s = 0.55$ cm^{-1}, $n = 1087$ 1^{-1}, $E = 5.6 \times 10^{-4}$ g/n, pH = 5.

cal absorption, because glucose oxidase is inactive below
pH 2. Fig. 1 shows that $\lfloor k_1^* a(1 - A_o/\tau \rfloor /C_{E_O}$ against A_o
for oxygen absorption with soluble glucose oxidase is linear
through the origin. The slope gave the specific rate $k = 5.3 \times 10^{-2} 1$/gsec. For arginate glucose oxidase, the plot
$(k_1^* a_s (1 - A)_o /\tau]$ versus A_o is independent of the stirred
speeds ranging from 240 to 640 r.p.m. as illustrated in Fig.
1 and thus the external diffusion around arginate glucose
oxidase were negligible. The value of $k'' a_s$ is obtained from
the slope.

In oxygen transfer for alumina glucose oxidase in a continuous stirred tank that is a liquid-solid system, the first
term in the left hand side in Eq. 6 is neglected. Fig. 2
reveals that the plot of $\tau/(1 - A_o)$ against $1/A_o$ is independent of the stirred speeds and is linear through origin.

Fig. 3 shows the plot of C_{E_o}/R or a_s/R against $1/C_{A_o}$
for crude and arginate invertase in a batch stirred tank in
various speeds. It was found that the former reaction oc-

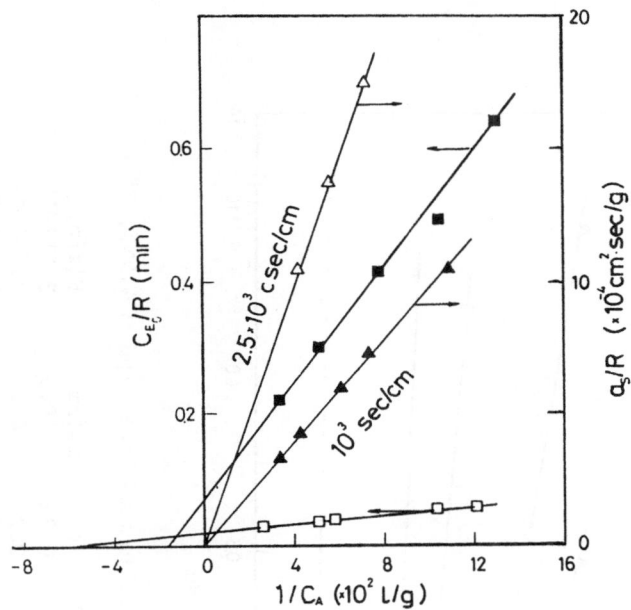

<u>Fig. 3</u>:   Sucrose transfer with crude and arginate invertase
          in a batch stirred tank.   (pH = 4.5, 25°C)
          □ crude invertase, C_{E_O} = 0.01~0.1 g/1, N = 300~800
          r.p.m.,
          △ arginate invertase, $\overline{d_p}$ = 0.31 cm, a_s = 0.36~0.61
          cm^{-1}, E = 3.3×10^{-4} g/n, $\overline{n}$ = 1042 1^{-1},
          ▲ arginate invertase, $\overline{d_p}$ = 0.31 cm, a_s = 0.032~0.13
          cm^{-1}, E = 3.5×10^{-3} g/n, $\overline{n}$ = 104 1^{-1},
          ■ arginate invertase, $\overline{d_p}$ = 0.043 cm, a_s = 0.036
          cm^{-1}, E = 1.3×10^{-5} g/n, $\overline{n}$ = 6316 1^{-1}.

cured in solid as same as the latter, because crude invertase was very poor soluble in water for one hour at 850 rpm.
Thus K_m' = 16.4g/1 and k' = $0.17sec^{-1}$.

    The apparent rate of sucrose for 3.1mm spheres or ar-
ginate invertase was not Michaelis-Menten type, but the
pseudo first order even through the high content of invertase. For 0.43 mm arginate invertase, it was found Michaelis-Menten type that k' = $0.046sec^{-1}$ and K_m' = 62.5g/1. This
is because the diffusion resistance in calcium arginate
membrane is very large in a big sphere. It is necessary to

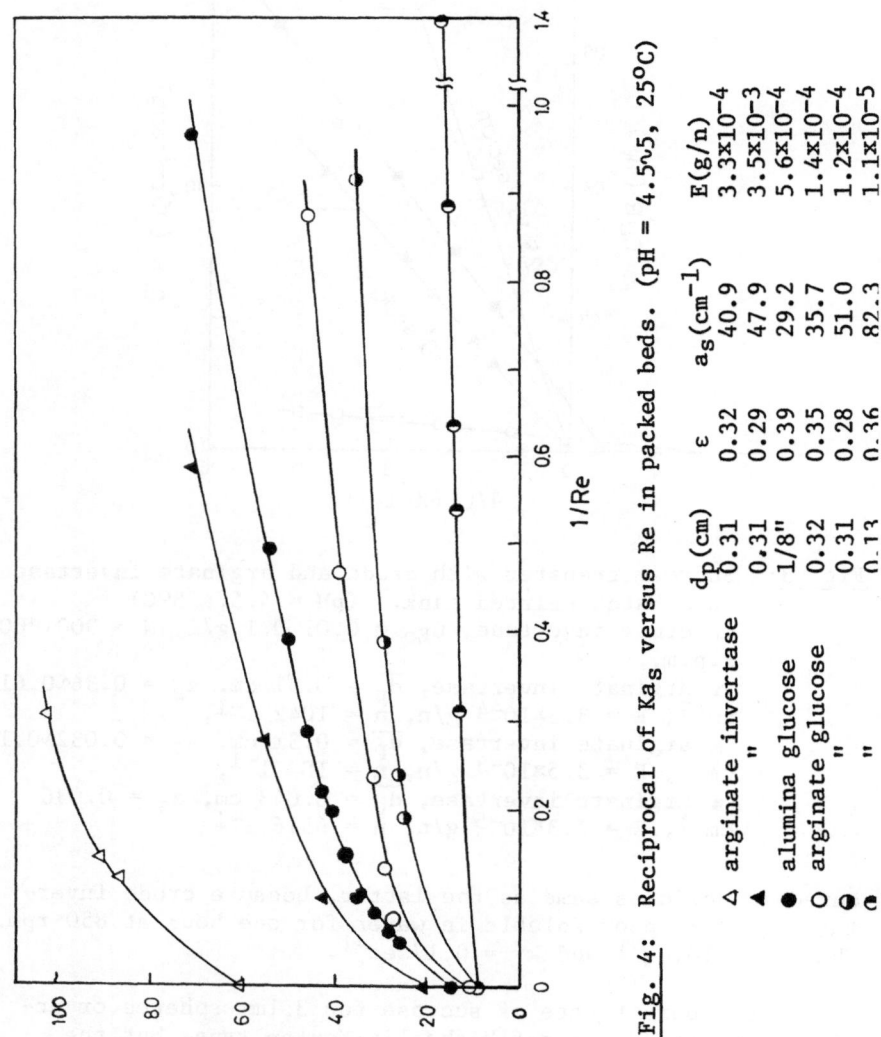

Fig. 4: Reciprocal of Ka_s versus Re in packed beds. (pH = 4.5∿5, 25°C)

	d_p(cm)	ε	a_s(cm⁻¹)	E(g/n)
△ arginate invertase	0.31	0.32	40.9	$3.3×10^{-4}$
"	0.31	0.29	47.9	$3.5×10^{-3}$
● alumina glucose	1/8"	0.39	29.2	$5.6×10^{-4}$
○ arginate glucose	0.32	0.35	35.7	$1.4×10^{-4}$
◐ "	0.31	0.28	51.0	$1.2×10^{-4}$
◑ "	0.13	0.36	82.3	$1.1×10^{-5}$

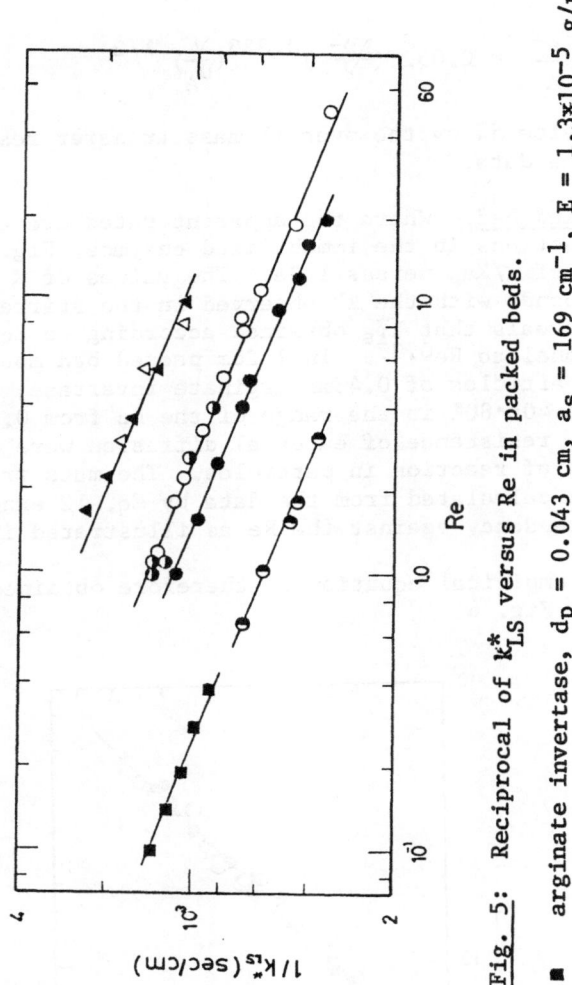

Fig. 5: Reciprocal of k_{LS}^* versus Re in packed beds.

■ arginate invertase, $d_P = 0.043$ cm, $a_S = 169$ cm^{-1}, $E = 1.3\times10^{-5}$ g/n.

investigate the effect of external diffusion on the overall
mass transfer rate in high concentration of sucrose. The
mass transfer coefficient in the liquid-solid stirred tank
can be estimated from the equation given by Traybal (13).

$$\frac{k_{1s}^{*}T}{D_A} = 0.052 \ (\frac{ND^2}{\nu})^{0.833}(\frac{\nu}{D_A}) \ 1/2 \tag{17}$$

It was below 5% ov the overall mass transfer resistance in
all of the data.

Packed Bed. Where the apparent rates are the first
order reactions in the immobilized enzymes, Fig. 4 shows
the plot of $1/Ka_s$ versus $1/Re$. The values of K at $1/Re = 0$ corresponds with the k" observed in the stirred tanks.
Fig. 5 reveals that k_{1s}^{*} obtained according to Eq. 14 is
proportional to $Re^{0.37}$. In 1.2cm packed bed mounted 9000
spheric particles of 0.43mm arginate invertase, the conversion were 40~80% in the range of the Re from 0.1 to 0.8
which the resistance of external diffusion were comparable
with that of reaction in particles. The mass transfer coefficient calculated from the data by Eq. 12 exhibits the
similar tendency against the Re as illustrated in Fig. 5.

The empirical equation is therefore obtained as illustrated in Fig. 6.

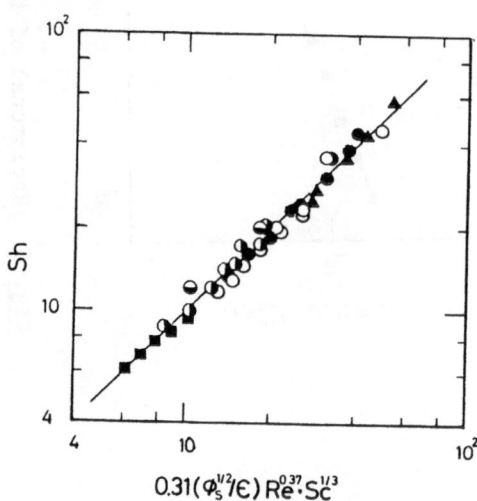

Fig. 6: Empirical equation.

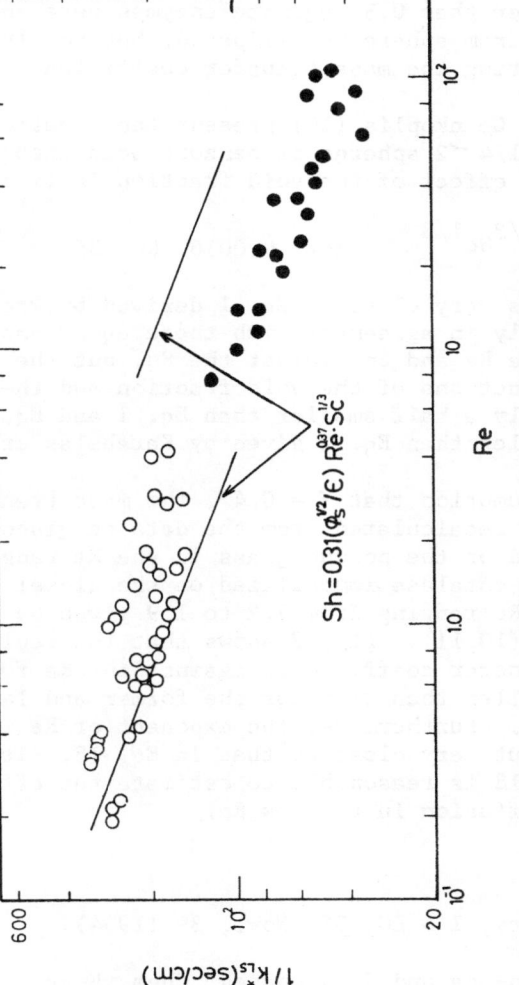

Fig. 7: Reciprocal of k_{LS}^* versus Re in packed beds.

○ Rovito & Kittell[10], glucose oxidase on porous glass (immobilized enzymes), $d_p = 0.035$ cm, $a_s = 207$ cm^{-1}, $D_A = 2.4 \times 10^{-5}$ cm^2/sec, $\mu = 0.01$ g/cm·sec,

● Traher & Kittell[12], catalase on Ni-SiO$_2$-Al$_2$O$_3$ (immobilized enzymes), $d_p = 0.06$ cm, $a_s = 121$ cm^{-1}, $D_A = 10^{-5}$ cm^2/sec, $\mu = 0.015$ g/cm·sec.

In the figure: $Sh = 0.31(\phi_s^{1/2}/\epsilon)$ Re$^{0.37}$·Sc$^{1/3}$

$$Sh = \frac{0.31}{\epsilon} \phi^{0.5} Re^{0.37} Sc^{1/3} \qquad for \quad 0.1 \quad Re \quad 50 \qquad (18)$$

where ϕ is the shape factor defined as surface area divided by the 2nd power of particle diameter. Where the void fraction ϵ was smaller than 0.3 arginate enzymes were observed to be deformed from sphere to ellipsoid, but Eq. 18 is capable of estimating the mass transfer coefficient.

Wilson and Geankoplis (16) present the equation from dissolution of 1/4 ~ 2"spheres of benzoic acid into water in addition to the effect of the void fraction in literature

$$Sh = \frac{1.09}{\epsilon} Re^{1/3} Sc^{1/3} \qquad for \quad 0.0016 \quad Re \quad 55 \qquad (19)$$

This equation is very close to Eq. 1 derived by Pfeffer(9). Eq. 18 is closely in agreement with these equations for the exponents of the Re and Sc against the Sh, but the constant included the functions of the void fraction and the shape factor is roughly a half smaller than Eq. 1 and Eq. 19 and one fourth smaller than Eq. 2 given by Karabelas et al (5).

On the assumption that $\epsilon = 0.47$, the mass transfer coefficients were recalculated from the data on glucose oxidase immobilized on the porous glass in the Re ranging from 0.2 to 4.2, and catalase immobilized on the nickel-silica alumina in the Re ranging from 7.2 to 119 given by Kittrell and colleagues (10,11). Fig. 7 shows that the reciprocal of the mass transfer coefficients against the Re for Eq.18 is somewhat smaller than that for the former and larger than that for latter. Furthermore, the exponent of Re is not equal to 0.5, but very close to that in Eq. 18. It indicates that Eq. 18 is reasonable to estimate the effect of the external diffusion in the low Re.

REFERENCES

1. J.J. Carberry, I & EC, 56, Nov., 39 (1964).

2. I. Colquhoun-Lee and J. Stepanek, Chem. Engrs., Feb.,
    108 (1974).

3. C.E. Dryden, D.A. Strang and A.E. Withrow, Chem. Eng.
    Prog. 49, 191 (1953).

4. B.J. Caffney and T.B. Drew, Ind. Eng. Chem., 42, 1120 (1950).

5. A.J. Karabelas, T.H. Wagner and T.J. Hanratty, Chem. Eng. Sci., 26, 1581 (1971).

6. Y.Y. Lee and G.Y. Tsao, Chem. Eng. Sci., 27, 1601 (1972).

7. L.K. McCune and R.H. Wilhelm, Ind. Eng. Chem. 41, 1124 (1949).

8. H. Obata, Y. Matsunaga and T. Tokuyama, Annual Meeting, Chem. Soc. Japan (April, 1974).

9. R. Pfeffer, I. & EC, Fundls, 3, 381 (1968).

10. B.J. Rovito and J.R. Kittrel, Biotechnol. Bioeng. 15, 143 (1973).

11. D. Thoenes and H. Kramers, Chem. Eng. Sci., 8, 271 (1958).

12. D. Traher and J.R. Kittrell, ibid., 16, 419 (1974).

13. R.T. Traybal, "Liquid Extraction" 2nd ed., McGraw-Hill Co., New York, N.Y. (1963).

14. D.W. vanKrevelen and P.J. Hoftijzer, Chem. Ind. XXIeme Congr. Int. Chim. Ind., p. 168 (1948).

15. J.E. Williamson, K.E. Bazaire and C.J. Geankoplis, I 7 EC, Fundls., 2, 126 (1963).

16. E. Wilson and C. Geankoplis, ibid., 5, 9 (1966).

NOMENCLATURE

A_{in} : dimensionless concentration of oxygen at inlet, C_{Ain}/C_{Ai}.

A_o : dimensionless concentration of oxygen at outlet or in the bulk of liquid, C_{Ao}/C_{Ai}.

a : interfacial area between gas and liquid per unit liquid volume.

a_s : interfacial area of particles immobilized enzymes per unit liquid volume.

C_A :   substrate concentration.

C_{Ai} :   oxygen concentration at interface between gas and liquid or saturated concentration of oxygen.

C_{Ao} :   substrate concentration at outlet or in the bulk of liquid.

C_{As} :   substrate concentration at interface of immobilized enzymes.

C_{Eo} :   enzyme weight per unit liquid volume.

C_G :   glucose concentration.

D :   impeller diameter.

D_A :   diffusivity of substrate.

d_p :   diameter of particle immobilized enzymes.

$\bar{d}^p$:   mean diameter of particles immobilized enzymes.

EP :   enzyme weight per a particle immobilized enzymes.

K :   apparent specific rate of immobilized enzymes defined by Eq. 14.

K_m :   Michaelis constant.

K'_m :   apparent Michaelis constant.

k :   specific rate.

k' :   apparent specific rate.

k'' :   apparent specific rate defined by Eq. 15.

k_1^* :   liquid film mass transfer coefficient between gas and liquid.

k_{1s}^* :   liquid film mass transfer coefficient between liquid and immobilized enzymes.

N :   stirred speed.

n :   number of parlicles immobilized enzymes per unit liquid volume.

Pe :   Pecletnumber, $ReSc$.

Re :   Reynolds number, $d_p v_z / \nu$.

Sc :   Schmidt number, ν / D_A.

Sh :   Sherwood number, $k_{1s}^* d_p / D_A$.

T :   Tank diameter.

V_1 :   liquid volume of reactors.

v_z :   linear velocity of liquid in longitudinal direction of packed bed.

Greek letters:

α $\dfrac{k' C_{Eo}}{k_{1s}^* a_s}$ as indicated in Eq. 10.

ϵ void fraction.

ν kinetic viscosity.

τ residence time of liquid.

ϕ shape factor.

CELLULASE PRODUCTION BY THERMOPHILIC ACTINOMYCES STRAIN

MJØr AND CHARACTERIZATION OF THE ENZYME

Tah-Mun Su

General Electric Research and Development Center

Schenectady, New York

The utilization of cellulose is one of the major challenges to technology. Cellulose is a photosynthetic product of green plants and is a replenishable resource. The annual production of cellulose has been estimated at 80 billion tons (1). However, most of the cellulose is not in a form suited to human needs as food and fuel. The cellulose occurs mainly in combination with other polymers such as lignin and hemicellulose as insoluble lignocellulose fibers. The efficient utilization of cellulose by conversion to other useful materials would be simplified if the cellulose were hydrolyzed into soluble sugar such as glucose (Fig. 1). Complete hydrolysis can be achieved in 1% sulfuric acid at an elevated temperature (170-180°C)(2). Unfortunately, the process has a low yield of glucose and produces toxic byproducts such as furfural and acetic acid. Cellulose can also be degraded by living organisms at a moderate temperature and pH. The organisms product extracellular enzymes (cellulases) that hydrolyze the cellulose into soluble sugar. The sugar is utilized for growth. A process based on enzymatic saccharification might involve the use of isolated enzymes or the enzyme produced in situ by the organisms growing in the culture. The development of a practical enzymatic saccharification system requires (a) identification of an efficient microbial source of enzyme and (b) characterization of enzyme generation and activity so as to define the optimal operating procedure.

Two biological sources of cellulase have received much interest in recent years; <u>Trichoderma</u> <u>viride</u> and thermo-

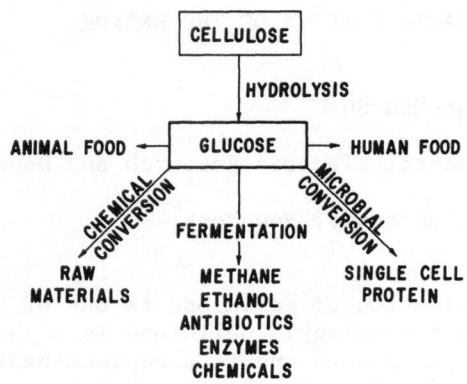

Fig. 1

philic actinomyces. T. viride has been shown to produce a
stable and active enzyme (3). This cellulase from T. viride
can hydrolyze about 37% of finely ground pure cellulose in
48 hours at 50° (3). This yields a 14% glucose syrup. However, this enzyme is not efficient in hydrolyzing the lignocellulose fibers. The more complete and rapid hydrolysis
requires substrate pretreatment i.e., ball-milling or
chemical swelling to increase the accessibility of the cellulose to the cellulase. The commercial development of an
enzymatic saccharification process based on T. viride is
currently limited by the slow rate of enzyme production and
the added expenses of pretreatment (4).

A strain of thermophilic actinomyces, MJ, was isolated
by W.D. Bellamy and a mutant, MJØr, resistant to high
temperature actinophage was selected by Dr. A.M. Chakrabar-
ty. MJØr has been found to digest lignocellulose fiber at
a rapid rate (1). This organism has been used to investigat
the possibility of producing single cell protein from agricultural wastes such as cow manure. Thermophilic conversion offers potential advantages of rate, pasteurization
and temperature regulation over the mesophilic and anaero-
bic fermentations. In this report, we will characterize
the cellulase activity responsible for the rapid cellulose
utilization of MJØr, compare the overall growth behavior
of MJØr with that of T. viride and characterize the acti-

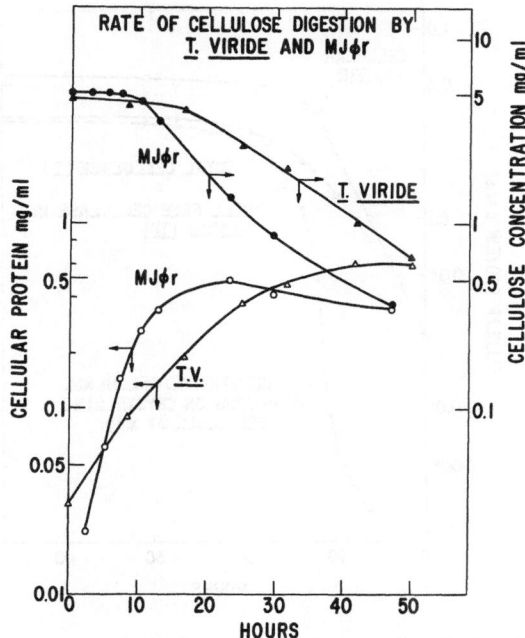

Fig. 2: Growth of Thermophilic Actinomyces Strain MJØr and
Trichoderma Viride on 0.5% Microcrystalline Cellulose in Minimum Basal Medium; MJØr Grown at 55° and
pH 7.5 and T. Viride Grown at 28° and pH 4.8.
(o) MJØr Cellular Protein; (△) T. Viride Cellular
Protein; (●) MJØr Cellulose; (▲) T. Viride Cellulose.

vity and properties of the soluble cellulase responsible
for cellulose saccharification in this system.

It was shown by Bellamy that MJØr has optimum growth
at 55° and pH 7.5-7.8 (1). In this study, MJØr was grown
in a fermentor containing 0.5% microcrystalline cellulose
(Avicel) in a minimum basal medium at constant pH (7.5) and
temperature (55°). The profiles of substrate depletion and
culture growth are shown in Fig. 2. A mycelial inoculum
was used to minimize the lag time during the fermentation.
T. viride was grown at constant pH (4.8) and temperature
(28°C) in a basal medium containing 0.5% Avicel (5) and a

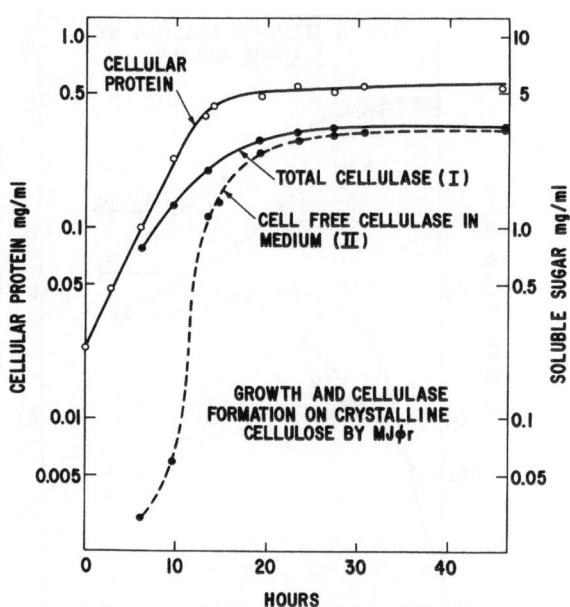

<u>Fig. 3</u>: Thermophilic Actinomyces Strain MJØr Grown on 0.5%
Microcrystalline Cellulose at 55° and pH 7.5. En-
zyme Activity Expressed as Mg/Ml of Soluble Sugar
Produced from Incubating 10% Avicel in Enzyme Solu-
tion at 55° and pH 7.5. (O) Cellular Protein;
(•–•) Total Cellulase; (•–•) Cellulase in Cell-Free
Filtrate.

mycelial inoculum was also used. Culture growth was followed by the increase in cellular protein as measured by the
Lowry method. Substrate depletion was followed by assaying
the undigested cellulose. Undigested cellulose obtained by
centrifugation was dissolved in 72% sulfuric acid and then
determined as glucose by the anthrone method. The cellulose depletion and the cell growth of MJØr were faster than
that of <u>T. viride</u>. These results suggest that either the
activity of MJØr cellulase or the rate of cellulase production or both are superior to that of <u>T. viride</u> under these
fermentation conditions.

The kinetics of cellulase production by MJØr are

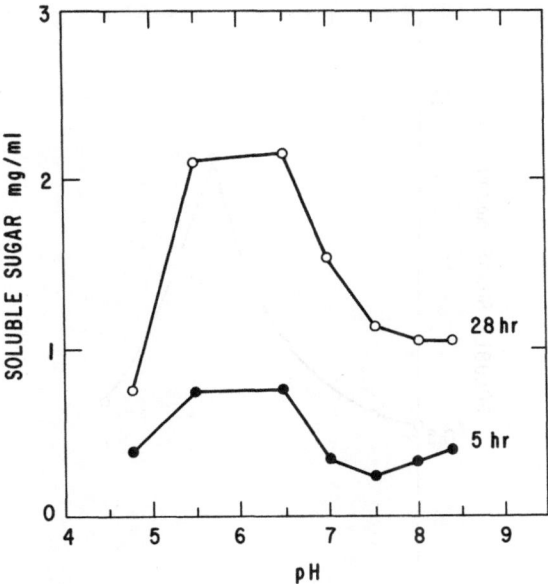

<u>Fig. 4</u>: pH Profile of Cellulase Activity from Thermophilic
Actinomyces Strain MJØr; (●) Enzyme and Substrate
Incubated for 5 hours; (○) Enzyme and Substrate
Incubated for 28 hours.

shown in Fig. 3. Curve I represents the enzyme activity
in fermentation broth which contains mycelial cells and
undigested cellulose. Curve II describes the enzyme activity in cell-free filtrate. The activity is expressed as
mg of soluble sugar produced by incubating the mixture of
1 ml of broth or filtrate and 1 ml of 20% Avicel in pH
7.5 buffer solution at 55°C for 18 hours. Merthiolate at
0.005% was added into the mixture to prevent microbial
growth. The soluble sugar was determined as glucose by
the anthrone method. Curve I shows that the total amount
of cellulase in the system roughly parallels the total
cellular protein. Maximum enzyme activity was obtained in
the stationary phase. No lag time was observed for either
the culture growth or the enzyme production. On the other
hand, a considerable lag time was observed for the appearance of cellulase in the cell-free filtrate as shown in

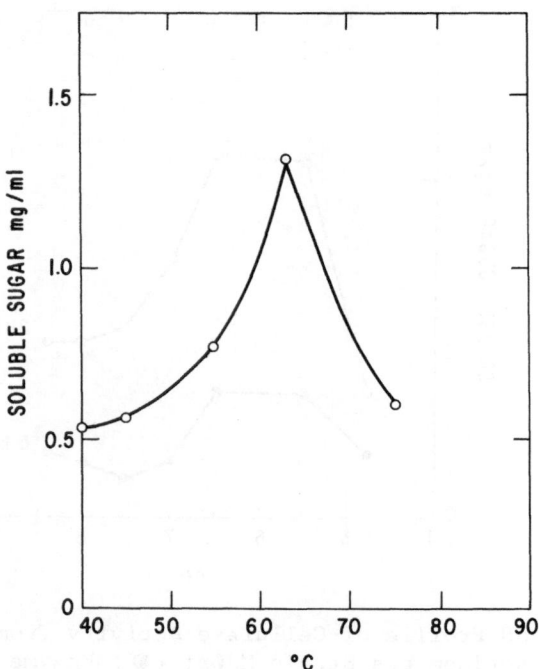

<u>Fig. 5</u>: Temperature Profile of Cellulase Activity From
         Thermophilic Actinomyces Strain MJØr; Activity
         Expressed as Mg/Ml Soluble Sugar Produced at pH
         6.0.

Curve II. Mandrels has reported that the <u>T. viride</u> cellulase is strongly adsorbed on the Avicel (6). Therefore, it
is likely that in the early stage of fermentation, the MJØr
cellulase also is adsorbed on the cellulose and hence can
be detected in the filtrate only after the surface of the
cellulose is saturated. This could explain the observed
lag time of enzyme appearance in the filtrate.

        The enzyme characterization was performed on crude
cell-free filtrate without purification. The pH and temperature profiles are shown in Fig. 4 and Fig. 5, respectively. To obtain the pH profile, the enzyme activity was
measured by incubating the mixture of 1.5 ml of 2% Avicel
in selected buffer solutions and 0.5 ml of cell-free fil-

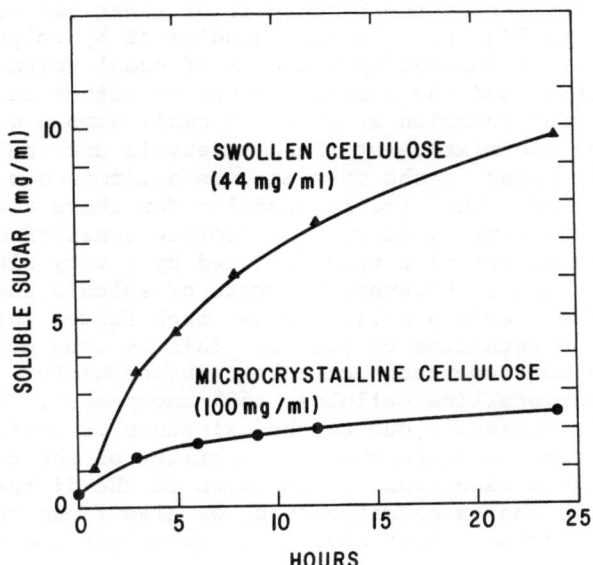

Fig. 6: Saccharification of Cellulose by Cell-Free Cellu-
         lase from MJØr. Substrate and Enzyme Mixture In-
         cubated at 55° and pH 7.5 and Soluble Sugar Mea-
         sured as Glucose. (●) Microcrystalline Cellulose,
         100 Mg/Ml; (O), Swollen Cellulose, 44 Mg/ml.

trate at 55° for 5 hours and 28 hours. The total soluble
sugar production after the incubation was determined as
glucose by the anthrone method. The optimum pH was at 6.2.
There was a suggestion of another maximum at elevated pH.
The temperature profile was measured at pH 6.0. The procedure was essentially the same one described for the pH profile except that the incubation time was 3 hours. The optimum temperature was found to be 63°C.

    The kinetics of cellulose saccharification were studied
with microcrystalline and swollen amorphous celluloses.
The microcrystalline cellulose is about 40% crystalline.
The swollen cellulose was prepared by dissolving the microcrystalline cellulose in cold 85% phosphoric acid, repre-
cipitating in water and washing repeatedly as described by
Walseth (7). It was used directly without drying. The

kinetic profiles of the hydrolysis of these two substrates
are shown in Fig. 6. The rate studies of hydrolysis were
carried out by incubating mixtures of equal volumes of cell-
free filtrate and the desired amount of either cellulose in
pH 7.5 buffer solution at 55°C. A small sample was with-
drawn from the mixture at timed intervals and the amount
of soluble sugar in the filtrate was determined by the anthrone method. The kinetic profiles for these two substrates were very similar. The profile consists of a very
rapid initial reaction rate followed by a very slow rate in
the later stage. However, the rate of soluble sugar production from swollen cellulose was much faster than that
from microcrystalline cellulose. This is true even though
there was more amorphous cellulose in the mixture containing
the microcrystalline cellulose (60% amorphous). This difference is primarily due to the existence of crystalline
regions which decrease the accessibility of the cellulose
for enzymatic reaction. In addition to the difference in
the rate of the saccharification, we also found that the
cell-free enzyme cannot completely hydrolyze the microcrystalline cellulose even with prolonged incubation. This
observation is in contrast to that seen in the fermentor
where the culture is able to utilize 90% of Avicel within
48 hours. One possible explanation for this discrepancy is
that the cellulase produced by MJØr is a multiple enzyme
system, and that during the saccharification process, components that are important to the hydrolysis of crystalline
cellulose are progressively deactivated. Since the culture
in the fermentor can continuously resupply the unstable
enzyme, the culture is able to hydrolyze the crystalline
cellulose completely. Furthermore, there is evidence of
progressive deactivation of components required to hydrolyze the swollen cellulose. As shown in Fig. 7, the initial rapid rate of hydrolysis gradually decreased after
three hours of incubation. However, the initial rapid rate
of hydrolysis could be maintained when the fresh enzyme was
repeatedly introduced to the mixture of partially hydrolyzed
swollen cellulose. Similar results have been observed with
10% Avicel. In this experiment swollen cellulose in the
cell-free filtrate was incubated at 55° in a buffer solution (pH 7.5) and the supernatant was removed every three-
hour intervals as shown by the arrow in Fig. 7. The residual cellulose was washed once with distilled water and
then incubated again with the same amount of fresh enzyme.

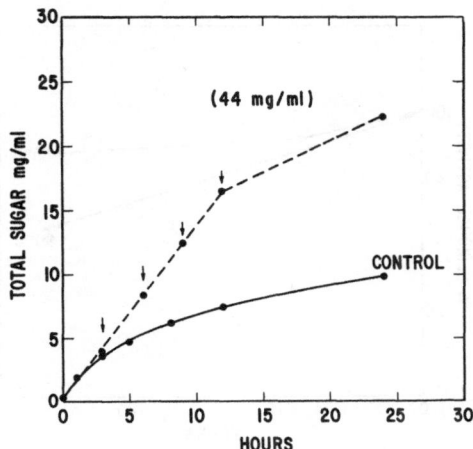

Fig. 7: Saccharification of Swollen Cellulose with Repeated
Addition of Fresh MJØr Cellulase at 55° and pH
7.5. Total Soluble Sugar as Glucose (●—●) Control;
(●–-●) Repeated Addition of Fresh Enzyme at Arrow
Shown.

        Thermal denaturation of protein is one of the pro-
cesses that could deactivate enzyme activity and decrease
the rate of saccharification. Therefore, we studied the
thermal stability of the cell-free cellulase at 55° and
60°C at constant pH (7.5). The cell-free filtrate was incubated with 10% Avicel as substrate in the constant temperature bath and the enzyme activity in the filtrate was
analyzed at timed intervals by the procedure described
above. As shown in Fig. 8, the cellulase is relatively
unstable at its optimum temperature (60°C) and exhibits a
first order decay with a half life of 24 hours. At 55°C
the enzyme is relatively stable. The enzyme loses about
20% of its original activity in the first 8 hours and only
decreases an additional 7% in the following 16 hours. The
existence of two different slopes of enzyme deactivation
at 55° indicates that the cellulase activity of MJØr is
probably dependent on two or more enzymes differing in
thermal stability.

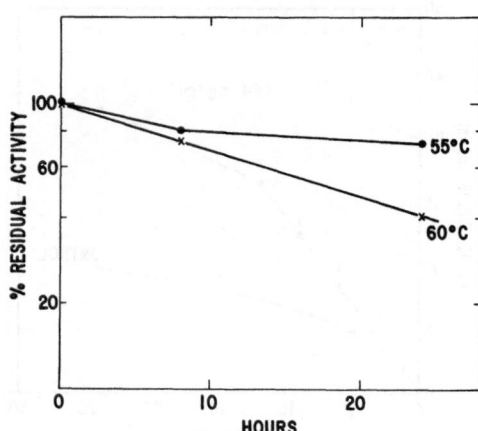

<u>Fig. 8</u>: Thermal Deactivation of Cellulase from Thermophilic
         Actinomyces Strain MJØr. Enzyme Solution Incubated
         at 50° or 60° at pH 7.5.

     From these experiments, we have learned that thermo-
philic actinomyces strain MJØr grows and degrades the microcrystalline cellulose faster than <u>T. viride</u>. The cellulase production by MJØr roughly parallels the cellular
protein and the cellulase activity reaches maximum activity
in the stationary phase. There is a lag time in the appearance of cellulase in the filtrate. The lag seems due to
the strong adsorption of the cellulase on the undigested
cellulose. The optimum pH and temperature of the cellulase
are near the optimum conditions for the growth of MJØr.
The enzyme hydrolyzes swollen cellulose to the greater extent than microcrystalline cellulose. With both substrates,
the enzyme gradually deteriorates during saccharification.
However, the initial fast rate of saccharification can be
maintained by repeated addition of fresh enzyme. The·cell-
free enzyme is relatively stable at 55° but a first order
decay with a half life of 24 hours is observed at 60°.
There are two slopes of decay at 55° which indicates that
the cellulase activity of MJØr is dependent upon enzymes
differing in thermal stability.

     We conclude that MJØr is a significant potential
source of cellulase to be used in the enzymatic sacchari-

fication process. However, the problem of enzyme deactivation will remain a barrier to economic utilization until future research provides the means to minimize the loss.

ACKNOWLEDGEMENTS

The author expresses his appreciation to Drs. J.A. Bergeron, R.E. Brooks and J.F. Brown, Jr. for helpful discussions and comments and to Miss I. Paulavicius for technical help.

REFERENCES

1. Bellamy, W., Biotech. and Bioeng. 16, 869 (1974).

2. Faith, W., Ind. Eng. Chem. 37, 9 (1945).

3. Mandel, M., Hontz, L. and Nystrom, J., Biotech. and Bioeng. 16, 1471 (1974).

4. Mandels, M., Microbial Sources of Cellulase, A paper presented at a Symposium on Cellulose as Sugar and Chemical Source, held at Univ. of Calif., Berkeley, Cal. (1974).

5. Mandels, M. and Weber, J., Cellulase and their application, Amer. Chem. Society, Advances in Chemistry Series 1969.

6. Mandels, M., Kostick, J. and Parizek, R., J. Polymer Sci., 36, 445 (1971).

ELECTROCHEMICAL PREPARATION OF ENZYME-COLLAGEN MEMBRANE AND ITS APPLICATION

Shuichi Suzuki, Masuo Aizawa and Isao Karube

Research Laboratory of Resources Utilization
Tokyo Institute of Technology
Ooakayma, Meguro-ku, Tokyo

The electrochemical preparation of membrane from fibrous protein such as collagen has been developed in our laboratory since 1966 (1,2,3).

Shaping of membrane by the electrochemical method consists of application of electrochemical properties of fibrous protein in a dilute suspension. It is possible to obtain anodic or cathodic deposits of considerable thickness with good physical (i.e., mechanical) and chemical properties. Articles such as tube, membrane and sack are obtained by using this method.

The electrochemical preparation of enzyme-collagen membrane has been proposed as one effective technique to immobilize enzymes without loss of enzymatic activity. This method offers considerable advantages of stability and continuous use of the enzyme.

The similar electrochemical complexation of enzymes in the collagen fibril network has also been studied by Vieth et al (4).

In this paper, the electrochemical preparation of enzyme-collagen membranes and their applications are summarized.

MATERIALS AND METHODS

Collagen. Calf skin was scraped to remove all adhering fat, muscle and scale. It was then minced finely with

253

a meat chopper in the ice. Globular proteins were extracted with a 10% sodium chloride solution as previously described by Cooper et al (5). The resulting collagen was washed sufficiently by suspending in cold distilled water. This collagen was then stored at -10°C.

Enzymes. Bacterial ɑ-amylase (10^5DNU/g) was obtained from Nagase Sangyo Co., Ltd. Urease (from Jack bean 4,000 unit/g) was obtained from Sigma Chemical Co. Catalase (from Beef Liver) was obtained from Tokyo Kasei Kogyo Co., Ltd. Uricase (from Candida utilis, 2.4 I.U./mg) was obtained from Oriental Yeast Co., Ltd. All solvents and reagents were analytical grade and deionized water was used in all procedures.

Suspension Preparation. The skin collagen was sufficiently swollen with an aqueous solution of hydrochloric acid or sodium hydroxide and finely divided into fibril suspension by mechanical shearing at 10 - 15°C. This suspension was adjusted to the required concentration and pH.

Apparatus. Figure 1 shows the apparatus for the electrochemical preparation of enzyme-collagen membrane. A cell of acrylic plastic 3 cm long, 3 cm wide and 11 cm deep, fitted with twin platinum anodes clipped to the sides and with a parallel platinum cathode placed centrally to produce an approximately uniform field was used. The surface area of each electrode was 4 x 2 cm^2. Direct current was supplied with a current stabilizer.

Preparation of Enzyme-Collagen Membrane. The solution for electrochemical preparation of the enzyme-collagen membrane contained from 50 to 100 ml of 0.45% collagen fibrils at pH 3.8 or 10.4 and aqueous enzyme solution (Collagen: Enzyme, 10:1). As salts prevented electrochemical forming of the membrane (6), enzyme solution was dialyzed against a large volume of distilled water for 24 hrs to remove salts in the enzyme sample. All subsequent operations were performed below 5°C. All experiments were carried out at constant current, with current density of 2 to 4 mA cm^{-2} for 2 mins without any circulation. The electrolyte was cooled in ice to avoid denaturation of enzymes. The wet membrane formed on the electrode was removed from the electrode and washed in a large volume of cooled water. After washing, enzyme-membranes were dried in a vacuum drying apparatus.

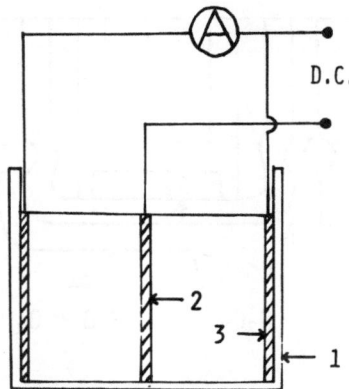

Fig. 1:   Apparatus for the electrochemical preparation of
          enzyme-collagen membrane.
          1. Cell $(3x3x11cm^3)$;   2. Cathode $(Pt:4x2cm^2)$;
          3. Anode $(Pt:4x2cm^2)$.

Enzyme Activity.   The urease activity of the membrane
was measured by the method of Van Slyke et al (7).
Amylase activity was measured by the method of Teller (8).
Amylose was obtained from Hayashibara Co., Ltd.   The enzyme
activity of catalase was determined as described by Chance
and Maehyl (9).   Uricase activity was measured by the method of Yamamoto and Nakagiri (10).

Assembly of the Hydrogen Peroxide Sensor.   The scheme
of the hydrogen peroxide sensor is illustrated in Figure 2.

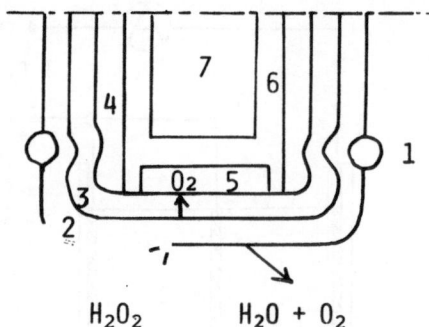

Fig. 2: Scheme of the bio-electrochemical sensor for hydrogen peroxide.
1. Rubber ring; 2. Catalase-collagen membrane;
3. Teflon; 4. Insulator; 5. Cathode(Pt);
6. Electrolyte (KOH); 7. Anode (Pt).

The sensor consists of a double membrane of which one layer
is catalase-collagen and the other is an oxygen-permeable
teflon membrane, an alkaline electrode, a platinum cathode
and a lead anode. The control solution was free from hydrogen peroxide and was saturated with dissolved oxygen. Both
the control and sample solutions were stirred magnetically
while measurements were taken.

RESULTS AND DISCUSSION

Effect of pH. Figure 3 shows the effect of pH on the
electrochemical preparation of collagen membrane. As is
evident, collagen membrane was formed on the cathode from
a pH of 2.5 to 5.3, and on the anode from a pH of 9 to 12.

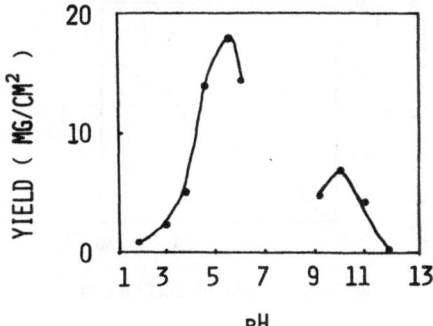

<u>Fig. 3</u>:   Effect of pH on electrochemical deposition of col-
              lagen. 50 ml of 1.5% collagen suspension was em-
              ployed. Current passed 0.53mAhr. 25°C.

In the alkaline condition, the collagen membrane slid off
the electrode as soon as the electrode was lifted from the
cell. Therefore, the amount of the aggregate could not be
determined.

    <u>Enzyme Content of Membrane (11)</u>. Figure 4 shows the
relationship between the amylase content of the collagen
suspension and that of the membrane prepared electrochemically. The amylase content of the membrane increased linearly with the increase in amylase content of the collagen
suspension. The ratio of the amylase to collagen in the
membrane was almost equal to that in the collagen fibril
suspension. Therefore, amylase may migrate to the cathode
with a mobility almost equal to that of the collagen fibrils.
Then amylase was codeposited on the surface of the cathode
with collagen fibrils. Figure 5 shows a flow sheet for

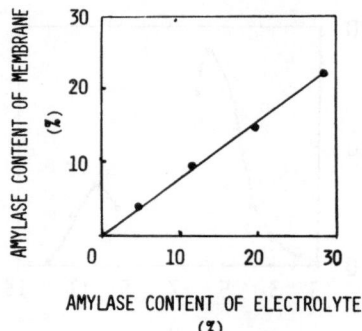

AMYLASE CONTENT OF ELECTROLYTE
(%)

Fig. 4:   Relationship between the amylase content of the
          collagen suspension and that of the membrane. A
          100 ml volume of 0.4% collagen suspension (pH 3.3)
          containing amylase was employed. 122V direct cur-
          rent was supplied through the cell for 2 min. at
          5°C.

the preparation of enzyme-collagen membrane which was established on the basis of the experimental results described
above.

     Activity of Enzyme-Membrane. Table 1 shows the rela-
tive activity of various enzyme-collagen membranes prepared
electrochemically. The highest relative activity of enzyme-collagen membranes was observed for catalase (Ca. 95%)
entrapped in collagen membrane. The example of collagen-
entrapped enzyme exhibiting low relative activity was alcohol dehydrogenase (Ca. 10%).

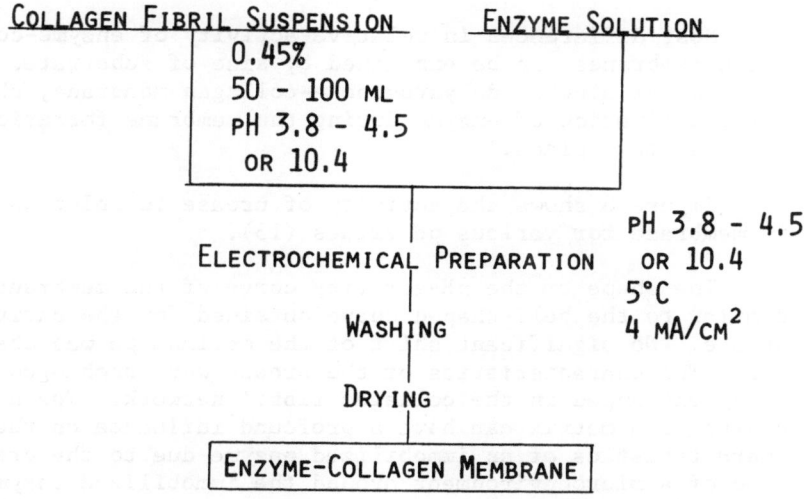

Fig. 5: Flow sheet for the preparation of enzyme-collagen membrane.

TABLE 1

RELATIVE ACTIVITY OF ENZYME-COLLAGEN MEMBRANE

Enzyme-Collagen Membranes	Relative Activity (%)
Amylase	55
Urease	51
Catalase	95
Uricase	43
Amp Deaminase	37
Alcohol Dehydrogenase	10

It is known that the relative activity of lattice entrapped enzymes is dependent on the nature and size of substrate and support (12).

Most differences in relative activity of enzyme-collagen membranes can be explained by size of substrate. In the case of alcohol dehydrogenase-collagen membrane, chemical inactivation of enzyme during the membrane formation may also take place.

Figure 6 shows the activity of urease in solution and in membrane for various pH values (13).

The shape of the pH-activity curve of the membrane was similar to the bell-shaped curve obtained for the native urease. No significant shift of the optimum pH was observed. The characteristics of the urease were unchanged by being entrapped in the collagen fibril network. The net charge of a matrix can have a profound influence on the characteristics of an immobilized enzyme due to the creation of a microenvironment around the immobilized enzyme (12). In general, the pH-activity curve of an enzyme bound to a negatively charge carrier (Seeing locally a higher hydrogen ion concentration) will be displaced toward more alkaline pH values. On the other hand, the pH-activity curve of an enzyme bound to a positively charged support will be displaced toward lower pH values.

It is known that the isoelectric point of collagen is between 7 and 8 (14). In the neutral condition, the net charge of collagen fibril is almost zero. The charges on the collagen fibrils may not influence the nature of the enzyme. This phenomenon can be seen commonly with other enzyme-collagen membranes.

The storage stability of collagen entrapped enzyme was examined. Amylase entrapped in collagen membrane retained its original activity after one week at 20°C, alcohol dehydrogenase lost no activity during one month of storage at 4°C.

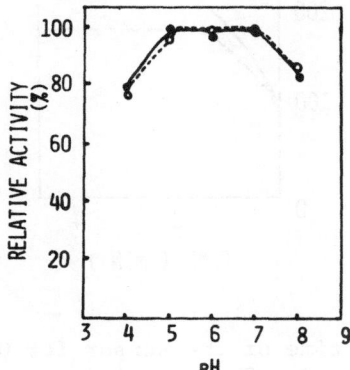

Fig. 6: Activities of urease at different pH values between
4 and 8. Activities were measured at 20°C in a 5
ml reaction mixture (pH 7.0, 0.2M phosphate buffer
containing 3% urea) with 0.5 mg urease or a piece
of membrane with equivalent enzymatic activity at
pH 7.0. ◯ Urease-collagen membrane. ●
Urease in solution.

Bio-Electrochemical Sensor (15). In Figure 7, the
response time of the hydrogen peroxide sensor is shown for
0.5 and 1.0 mmol of hydrogen peroxide per liter at 20°C.
The sample solution was first saturated with oxygen gas.
Saturated dissolved oxygen was responsible for the current
at time 0 in the sample solution.

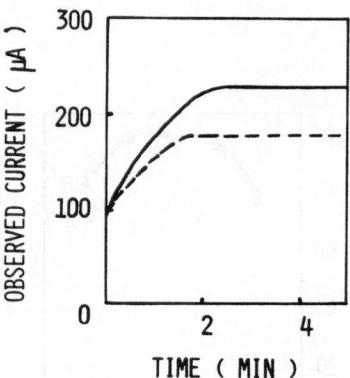

<u>Fig. 7</u>:   Response time of the sensor for 0.5 mmol 1^{-1}
         (----) and 1.0 mmol 1^{-1} (———) of hydrogen
         peroxide.

Hydrogen peroxide was decomposed to oxygen and water by
catalase

$$H_2O_2 \quad \underline{\text{catalase}} \quad H_2O + 1/2\ O_2$$

entrapped in the membrane, when the sensor was inserted in
the sample solution. Generation of oxygen by the decomposition of hydrogen peroxide caused oversaturation of dissolved
oxygen around the membrane, which increased the output of
the sensor. The output increased markedly with time, until
the steady state was reached. Although the response time
depends on the thickness of the membrane, enzyme concentra-

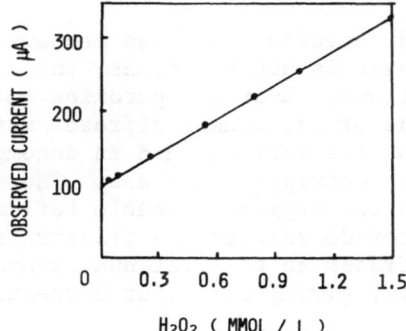

Fig. 8:   Calibration curve of the hydrogen peroxide sensor
over the concentration range 0 - 1.5 mmol 1^{-1} at
pH 6.2

tion and temperature, the steady state value was attained
within 1.5 minutes for 0.5 mmol per liter and 2 min. for
1.0 mmol per liter for hydrogen peroxide at 20°C. The current output related to the hydrogen peroxide concentration
is shown in Figure 8 in the concentration range 0 - 1.5 mmol
per liter. The current output is defined as the steady current at 3 min. after the insertion of the sensor. The oxygen generated by the hydrogen peroxide decomposition is responsible for the current, I-Io, where I is the observed
current for the sample solution and Io is the current for
the saturated dissolved oxygen. When the dissolved oxygen
in the sample solution was displaced with nitrogen gas, the
current output could be attributed solely to the generated
oxygen in the membrane. In the hydrogen peroxide concentration range above 1.5 mmol per liter, the current-concen-

tration plot curved toward the concentration axis.

Although the precise detection mechanism of the hydrogen peroxide sensor is not yet clear, the mechanism may be
explained as follows. Hydrogen peroxide near the catalase-
collagen membrane of the sensor diffuses into the membrane
or is adsorbed on its surface, and is decomposed to water
and oxygen by the entrapped catalase. The generated oxygen
diffuses to both the oxygen-permeable teflon membrane and
the hydrogen peroxide solution as illustrated in Figure 2.
Oxygen which diffuses to the platinum cathode is reduced
electrochemically, giving an output current.

On the other hand, uricase-collagen membrane was used
for uric acid determination (16). Figure 9 shows the relationship

between current output and uric acid concentration. Oxidation of uric acid was catalyzed by uricase entrapped in collagen membrane. Consumption of oxygen by the oxidation of
uric acid caused decrease of dissolved oxygen around the
membrane which decreased the current output of the sensor.
Saturated dissolved oxygen was responsible for the current
at time 0.

The new bio-electrochemical sensor is highly selective
and has many promising merits: 1.) a convenient and direct
measurement giving an electric signal, 2.) a quick response,
3.) applicability to coloured or opaque solutions, 4.) repeated use of the analytical reagent (enzymes), and 5.) repeated use of the sample solution because only a trace
amount of sample around the sensor is decomposed.

Recently we investigated biochemical fuel cells in
which the electrode reactions at the anode or the cathode
are catalyzed by enzymes (17).

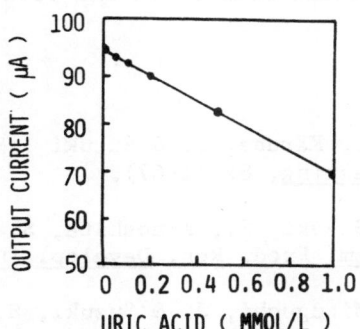

<u>Fig. 9</u>:   Calibration curve of the uric acid sensor over the
            concentration range 0 - 1.0 mmol 1⁻¹.

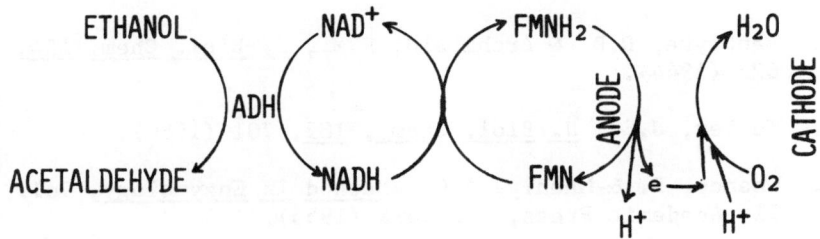

ADH : ALCOHOL DEHYDROGENASE

<u>Fig. 10</u>:   Biochemical fuel cell.

Figure 10 shows the scheme of the biochemical fuel cell of the alcohol dehydrogenase-collagen membrane. We are also investigating the lactate dehydrogenase and the invertase-glucose oxidase system. We are planning to use these cells as bio-electrochemical sensors in the food industry and the medical field.

REFERENCES

1. Mizuguchi, J., Karube, I. & Suzuki , S., Chem. Soc. Japan UEDA Meeting, 82 (1967).

2. Karube, I., Suzuki, S., Kinoshita, S. & Mizuguchi, J., Ind. Eng. Chem. Prod. Res. Develop. 10, 160 (1971).

3. Karube, I., Mizuguchi, J. & Suzuki, S., Kogyo Kagaku Zasshi, 74, 971 (1971).

4. Vieth, W.R. & Venkatasubramanian, K., Chem. Tech. 47, (1974).

5. Cooper, D.R., Russell, A.E., & Davidson, R.J., J. Amer. Leather Chem. Ass. 61, 423 (1967).

6. Karube, I. & Suzuki, S., Kogyo Kagaku Zasshi 74, 1267 (1971).

7. VanSlyke, D.D. & Archibald, R.M., J. Biol. Chem. 154, 623 (1944).

8. Teller, J.S., J. Biol. Chem., 185, 701 (1950).

9. Chance, B. & Maehly, A.C., Method in Enzymology, Vol. II, Academic Press, New York (1955).

10. Nakagiri, Y. & Yamamoto, T., Eiseikensa 20, 175 (1971).

11. Suzuki, S., Karube, I., & Watanabe, Y. Proc. IV IFS: Ferment. Technol. Today 375, (1972).

12. Zaborsky, O.R., Immobilized Enzymes 83, C.R.C. Press, (1973).

13. Karube, I. & Suzuki, S., Biochem. Biophys. Res. Commun. 47, 51 (1972).

14. Mizuguchi, J., Karube, I., & Suzuki, S., Kogyo Kagaku
     Zasshi 73, 2123 (1970).

15. Aizawa, M., Karube, I. & Suzuki, S., Anal. Chim Acta,
     69, 431 (1974).

16. Suzuki, S., Sonobe, N., Karube, I., & Aizawa, M., Chem.
     Letters 9 (1974).

17. Mizuguchi, J., Suzuki, S. & Takahashi, F., Bull. Tokyo
     Inst. Tech. 78, 27 (1966).

SCALE-UP STUDIES ON IMMOBILIZED, PURIFIED GLUCOAMYLASE, COVALENTLY COUPLED TO POROUS CERAMIC SUPPORT

H.H. Weetall, W.P. Vann, W.H. Pitcher, Jr.
Corning Glass Works, Corning, New York
D.D. Lee
Iowa State University, Ames, Iowa
Y.Y. Lee
Auburn University, Auburn, Alabama
G.T. Tsao
Purdue University, Lafayette, Indiana

ABSTRACT:

This report describes the development, scale-up, and economic characterization of an immobilized enzyme system for the hydrolysis of previously thinned cornstarch to high glucose syrups.

The results of these studies indicate that enzymes can be successfully immobilized on porous inorganic ceramic supports. The results also indicate that an immobilized glucoamylase column can be successfully operated at 40°C without bacterial contamination. It was also found that if contamination does occur, resanitization is achievable with little difficulty. The economic evaluation presented in this report was calculated based on the actual scale-up data presented.

Choice of Carrier

The economic production of dextrose from corn syrup, utilizing glucoamylase (GA) immobilized on controlled pore glass, has at least two serious disadvantages: (1) high cost of the carrier and, (2) dissolution of glass with time (1). The development of oxide coated glass has increased durability and half-life, which are important economic advantages. However, the cost of the carrier still exceeds the value of the product.

269

This report describes half-life studies of six porous ceramic carriers covalently bound (2,3) to purified GA. All of the carriers are less expensive to produce than porous glass and could make a presently uneconomic process more attractive. One of the carriers was utilized for a scale up to a one cubic foot column. Processing cost estimates based on the performance of this scaled-up column were made.

MATERIALS & METHODS

Support Materials

Physical and chemical data for the six support materials used in this study are summarized in Table I.

Alkylamine Derivative Preparation

Fifty g of support material was added to enough 10% γ-aminopropyltriethoxysilane (Union Carbide A1100 Silane) in distilled water so that the carrier was covered. The mixture was then adjusted to pH 3.45 with a 25% solution of NaOH. The mixture was reacted in a 75°C water bath for 3 hours, then washed with 4 volumes of distilled water and oven dried at 120°C for 15 hours.

Glucoamylase

Crude GA was obtained from NOVO Industries A/S, Copenhagen, Denmark and purified using isopropanol precipitation techniques. Activity of the purified enzyme was 65280 IU/g. One unit represents the production of 1μ mole of dextrose/min at 60°C.

Preparation of Immobilized Glucoamylase

GA was immobilized onto carrier using Schiff base coupling (4). The alkylamine derivative (usually 30-50 g) was added to a sufficient volume of 2.5% glutaraldehyde in 0.1M phosphate buffer at pH 7 so that the derivatized support material was covered. The reaction was performed at room temperature for 1 hour which included 30-45 minutes in a vacuum desiccator with occasional stirring. The carrier was then washed with 3-4 1 of distilled water. For immobilization, 100 mg of purified enzyme was offered per g of deriva-

TABLE I

PHYSICAL AND CHEMICAL PARAMETERS OF SUPPORT MATERIALS

Designation	Mesh	Composition		Pore Dia. Range (A)	Pore Dia. MP (A)	Pore Vol. (cm^3/g)
A	30/45	SiO$_2$ TiO$_2$	75% 25%	875 - 205	465	0.76
B	30/60	SiO$_2$ ZrO$_2$	90% 10%	700 - 185	435	0.76
C	30/60	SiO$_2$	100%	700 - 185	435	0.76
D	30/60	SiO$_2$ ZrO$_2$	84.3% 15.7%	575 - 110	235	1.30
E	30/60	SiO$_2$ Al$_2$O$_3$	75% 25%	575 - 205	435	0.89
F	30/45	TiO$_2$ MgO	98% 2%	500 - 205	410	0.53

tized support material. The enzyme was dissolved in cold
0.1M phosphate buffer at pH 7 and added to the support material in an ice bath with occasional stirring. The solution pH was maintained at 6.8-7.0 and the reaction continued for 2 hours. The immobilized enzyme (IME) was washed
with 3-4 1 of distilled water and stored at 4°C as a wet
cake.

Determination of Immobilized GA Activity

Initial immobilized GA activity was assayed at 60°C
for 1 hour at pH 4.5. Substrate was 50 ml 30% corn syrup
solids, prepared from 22 Dextrose Equivalent (D.E.) spray-
dried acid-enzyme thinned cornstarch (A.E. Staley Company,
Decatur, Illinois, to which a known quantity of enzyme was
added. Total glucose was determined with a Glucostat®
Special Kit (Worthington Biochemical Corp., Freehold, New
Jersey).

Substrate for Columns

Substrate used for this study was the same as for the
assay. The substrate feed pH was 4.5. Determinations of
D.E. were performed on the substrate whenever a new bag of
material was employed.

Immobilized Glucoamylase Columns

With the exception of support material B (Table I),
10 g wet weight (4.4 - 5.9 g dry depending on support material) of each immobilized enzyme was put into 3 water
jacketed columns at 50, 60 and 65°C. The 65°C column for
support B was omitted due to a limited supply of carrier.
Each column was assayed once each day for glucose produced,
with the Glucostat® Special Kit. Column flow rates were
adjusted where possible to yield initially about 70% or
less glucose production. Half-life determinations and regression analysis plots were performed on column data with
a model 10 Hewlett Packard calculator program.

When half-life determinations were completed, column
flow rate was set as low as possible, and Dextrose Equivalent (DE) determinations performed on the products.

TABLE II

COLUMN HALF-LIFE RESULTS

Support Material	Initial Act.[a] IME IU/g	T(°C)	Days Oper.	Half-Life (Days) $t_{\frac{1}{2}}$	LCL[b]	UCL[b]	Maximum DE Observed at 50°C	Deactivation Energy (Kcal/g mole)
A	1750	50	40	46.6	31.8	86.9	89.7	32.4
		60	19	8.8	7.6	10.5		
		65	14	5.6	4.6	7.1		
B	658	50	31	53.3	38.3	87.4	92.0	26.7[c]
		60	14	15.9	10.9	29.0		
		65	--	--	--	--		
C	2400	50	59	51.2	40.0	71.4	88.9	41.4
		60	16	13.8	10.1	22.0		
		65	9	3.4	2.7	4.7		
D	423	50	31	77.1	46.9	215.9	79.9	35.6
		60	12	11.2	9.7	13.3		
		65	12	7.5	5.9	10.4		
E	1729	50	59	113.3	76.5	217.6	89.6	53.9
		60	16	16.2	12.7	22.4		
		65	9	4.2	3.4	5.4		

[a] One unit of activity equals the production of 1.0μ moles of dextrose/min at 60°C.
[b] 95% confidence limits
[c] Calculated between 50-60°C, all others 50-65°C.

RESULTS AND CONCLUSIONS

Initial Activity of Immobilized Supports

Tables I and II summarize physical and chemical data and column, half-life results of the various support materials. Activities ranged from 423-2400 IU/g IME on a dry weight basis. Supports B and D had the lowest activity. Pore exclusion of enzyme would account for the low activity of support D (pore dia. 235 Å) but not support B (pore dia. 435 Å). Perchloric acid titration of silanized, zirconia-coated supports has shown less reactive amine groups present when compared to non-zirconia-coated supports. Also, reaction of β-naphthol with diazotized zirconia-coated arylamine support material indicated less color development or fewer diazonium sites available for coupling. Both B and D were zirconia coated, which probably contributed to the low activity of both supports.

Dextrose Equivalent Determinations at 50°C

Table II summarizes maximum DE observed for columns operated at 50°C. Substrate DE was actually 24.8. Supports A,B,C,E and F had product DE's of 87.1 - 92.0. Product DE for support D was 79.9. This value was probably due to the low initial activity and/or less than optimum flow rate for this material. It should be stressed that DE values may or may not represent maximum values. Pumps were set to provide a minimum flow rate which varies considerably from pump to pump. No attempt was made to optimize flow rate vs DE.

Half-lives of Immobilized Preparations

Half-life data are summarized in Table II. Figs. 1-6 are computer calculated regression analysis plots of 50,60 and 65° column half-life data for each support material. Columns operated at 50°C had appreciably longer half-lives than columns operated at 60°C and 65°C. Longest half-life obtained was 113 days with support E. Half-lives of the remaining columns operated at 50°C were ≤ 68% (46.6 - 77.1 days) of the longest half-life obtained.

The maximum DE achieved did not exceed 92. The use of an acid-enzyme spray-dried substrate may be the causative factor in the inability to achieve DE's of greater than

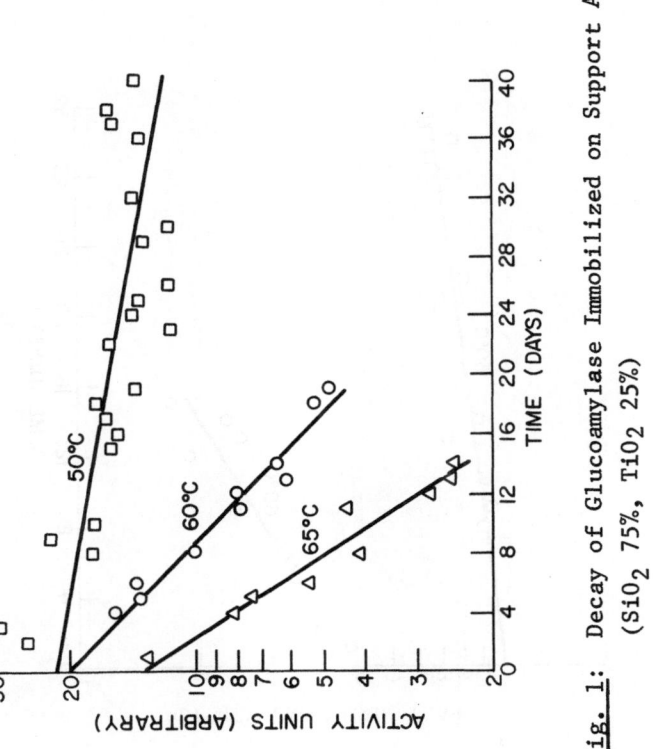

Fig. 1: Decay of Glucoamylase Immobilized on Support A (SiO$_2$ 75%, TiO$_2$ 25%)

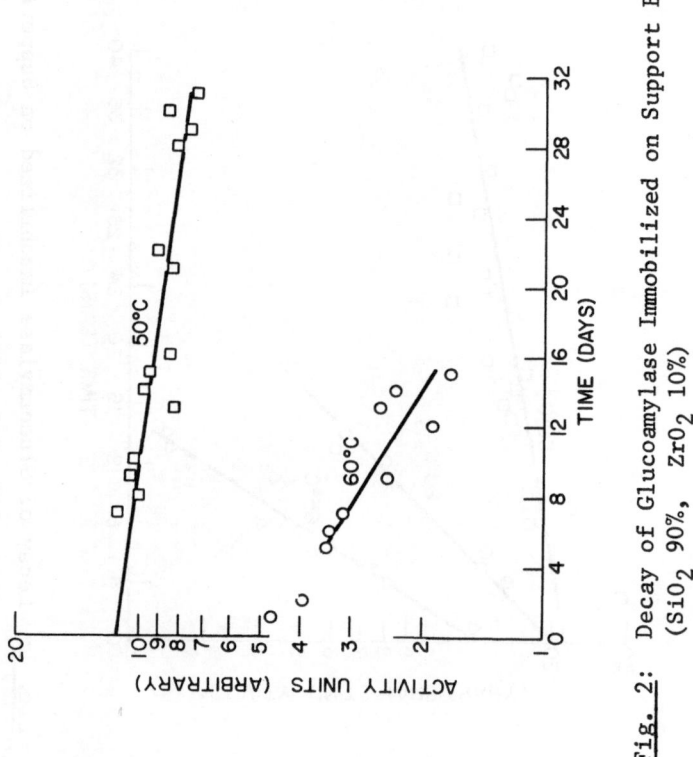

Fig. 2:   Decay of Glucoamylase Immobilized on Support B
          (SiO_2 90%, ZrO_2 10%)

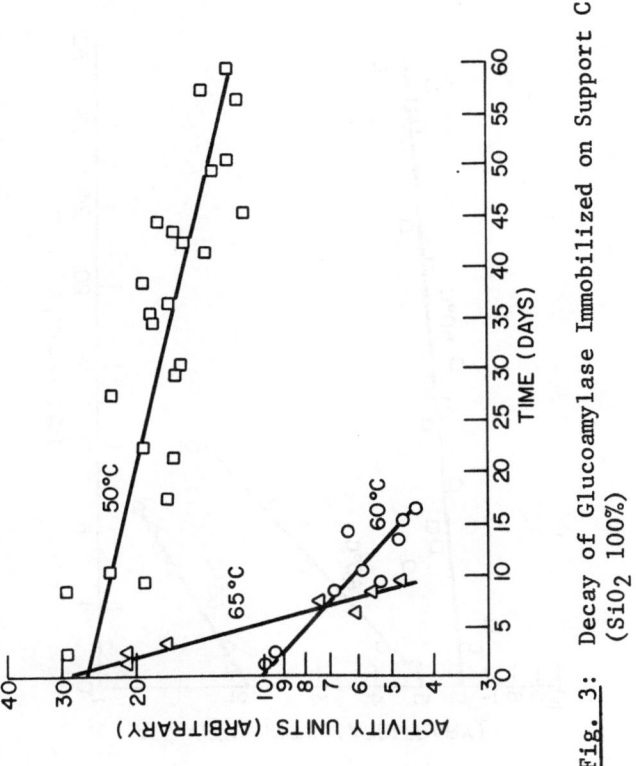

Fig. 3: Decay of Glucoamylase Immobilized on Support C
(SiO_2 100%)

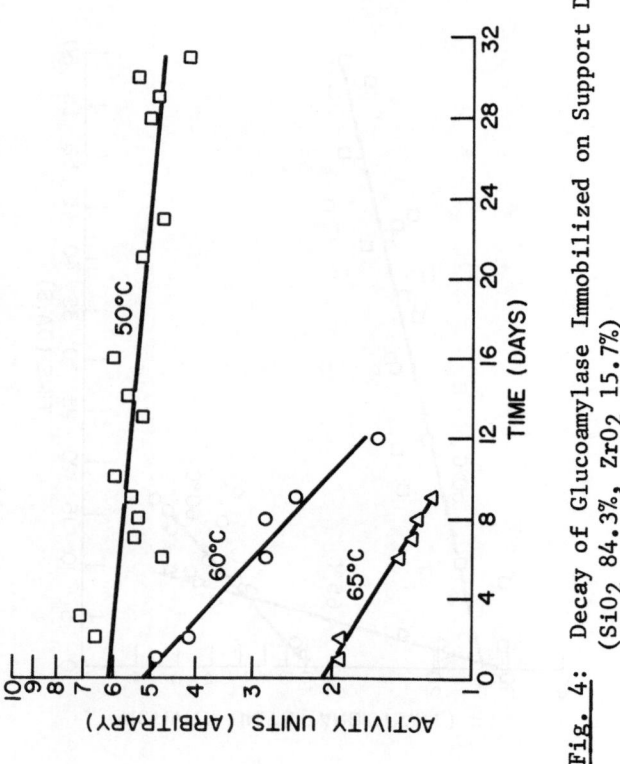

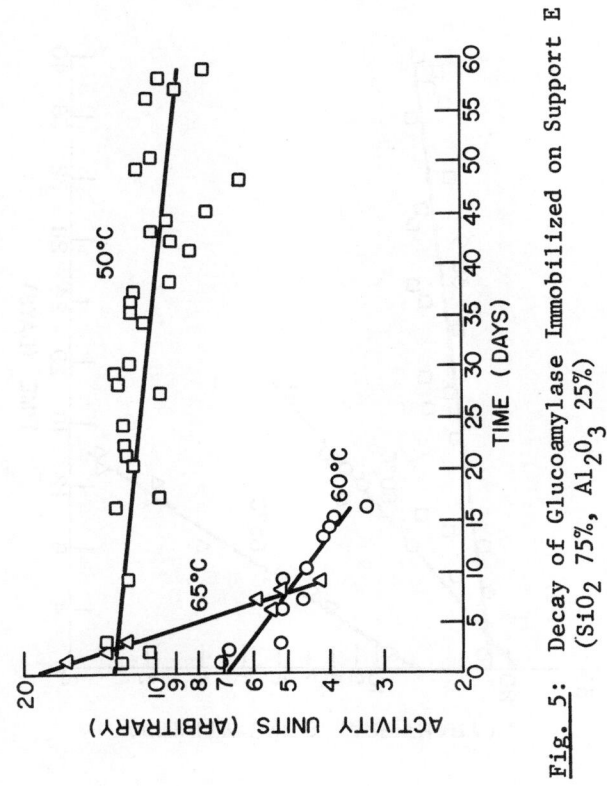

Fig. 5: Decay of Glucoamylase Immobilized on Support E
(SiO_2 75%, Al_2O_3 25%)

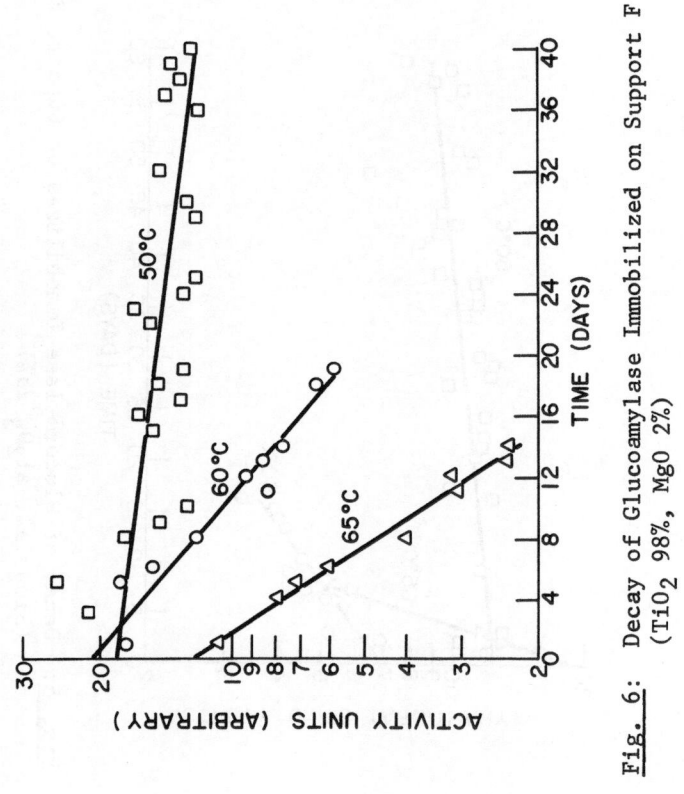

<u>Fig. 6</u>:   Decay of Glucoamylase Immobilized on Support F
             (TiO$_2$ 98%, MgO 2%)

95. This is borne out by the fact that GA studies on porous glass columns with freshly prepared enzyme-thinned cornstarch have given DE values in excess of 95 in the past.
Studies are in preparation to examine the substrate problem in greater detail.

A large batch of ceramic support material designation
C was chosen for the scale-up at Iowa State University.

This section of the report summarizes data collected
at the Iowa State University pilot plant for the scale-up
of immobilized glucoamylase. The design of the plant is
illustrated on Figure 7. For these studies only a single
reactor was used.

The scale-up began when approximately 32 lbs of SiO_2
carrier, previously silanized at Corning Glass Works was
placed in the column and coupled in situ. The coupling
data are presented in Table III. The entire process was
completed in less than 10 hours.

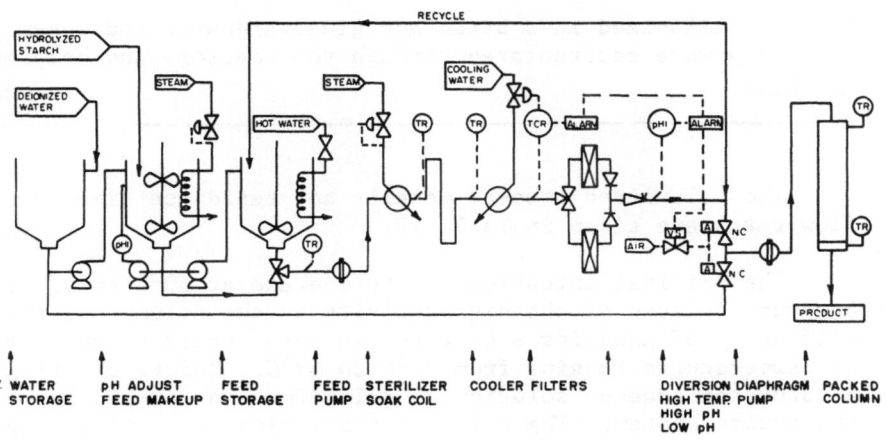

Fig. 7: Process Flow Diagram of Iowa State Glucoamylase
          Pilot Plant.

TABLE III

SCALE-UP IMMOBILIZATION DETAILS

Enzyme Support: SiO_2, 30-45 mesh, 400 $\overset{o}{A}$ Pore

Enzyme: Glucoamylase, Novo
         60,000 units/g   1 unit = 1μ mole of glucose/min

Enzyme Offered: 4 lbs to 31.8 lbs of carrier
                 (1 cubic feet bulk volume)

 Bulk Enzyme Activity Before Immobilization = 12,700 units/
                                                         ml
 Bulk Enzyme Activity After Immobilization = 1,400 units/
                                                       ml

Enzyme Attached: 3.56 lbs/31.8 lbs of carrier
                            or
                 112 mg/g carrier

Immobilized Enzyme Activity: 3,000 units/g carrier

Enzyme Bonding Method: Silanization of SiO_2 followed by
                        Glutaraldehyde Coupling of Enzyme

SiO_2 was silanized in a batch and glutaraldehyde and enzyme
solutions were recirculated through the reactor, and coupled
in situ.

The calculated process details and residence times vs
flow rates are given in Table IV.

The original intention was to operate at 50°C to 55°C.
However, because of the high activity of the column and the
difficulty of handling substrate, we were forced to operate
at temperatures ranging from 38°C to 42°C. Before starting,
a saturated aqueous solution of chloroform was pumped into
the entire system. The chloroform solution was replaced by
the sterile substrate as it passed into the column.

PILOT PLANT OPERATION:

The initial operation of the pilot plant was conducted

TABLE IV

PROCESS DETAILS

Reactor: Dimension = 6" I.D. x 5' H.

Volume = 0.98 CF

Estimated Reactor Void Volume with Packing = 0.3 CF (Pore Volume Excluded)

Reactor Residence Time

Production Rate (lbs Glucose/day)	Flow Rate (GPM)	*Nominal Residence Time (min.)	**Real Residence Time (min.)
1,200	0.30	24.4	7.48
1,000	0.25	29.3	8.99
800	0.20	36.6	11.22
500	0.13	56.4	17.30

* Reactor Volume divided by Flow Rate

**Reactor Void Volume divided by Flow Rate

---

with a starch hydrolysate of 24 D.E. called STAR-DRI 24-R, which is a partially acid hydrolyzed cornstarch. This was purchased from A.E. Staley. Feed solutions were made up to 30% by weight and the pH adjusted to approximately 4.5 by using hydrochloric acid. The initial production rate was about 1200 lbs/day dry weight glucose at a volumetric flow rate of 1100 ml/min through the enzyme column. Flow was from top to bottom, and the temperature was maintained at 37°C at the inlet with an outlet temperature of 38°C to 39°C.

The analysis of the product showed a glucose production of 89.6 to 90.0% by weight of the solids produced and a dextrose equivalent D.E. of 92.4 to 93.0. The enzyme activity during the 70 days of operation showed no detectable change when compared to the original activity. The values obtained for the glucose yield are lower than desirable (discussions with corn processors indicate that desired glucose levels

would be 95%.   However, most would settle for 94% glucose.)
from an industrial standpoint.   Equipment is being purchased and installed to enable the production of glucose
using Pearl starch hydrolyzed in a jet-cooker with   -amy-
lase.   Hopefully, this will significantly decrease the retrogradation and formation of non-hydrolyzable products
that now reduce the glucose yield.   Batch data with both
free enzyme and immobilized enzyme showed that a maximum
of 90-91% glucose could be obtained using the Staley 24
D.E. substrate.

The biological contamination of the enzyme column and
the product stream was measured by taking plate counts of
the product stream for bacterial and mold contamination.
After 30 days operation, the product stream showed a bacterial count of 36/ml, recycle stream 93/ml, and feed 220/
ml with no mold.   At 45 days the count was 6/ml in the product, 25/ml in the recycle, and 25/ml in the feed.   At 70
days the bacteria count was 1600/ml in the product, 4400/ml
in the recycle, and 20/ml in the feed.   The increase followed the introduction of unsterilized air into the column
when the feed pump was blocked.   Five days later, the count
was 41,000/ml in the product (Table V).

TABLE V

BACTERIAL CONTAMINATION

Operation Time	Bacterial Count/ml		Fluid
	Feed	Recycle	Product
30 days	220	93	36
45 days	25	25	25
70 days*	20	4,400	1,600
75 days			41,000
After Flush			300

*After column was contaminated with air.

No molds were detected at any time.   After an additional
three weeks of storage, column was operated for an additional 12 hrs. with no increase in count.

The column was washed with water by using bottom to
top flow for one day and this was followed by a wash with
a saturated chloroform solution through the entire system
including the column for three hours. The chloroform solution was pumped out with sterile feed. After four days,
the count was 300/ml in the product. The column was washed
with a chloroform solution and placed in the cold room at
4°C for 3 weeks. The plant was started up at that time and
run for 12 hours with the bacteria count at 4 hours being
330/ml of product, and 910/ml of product at 10 hours. The
column was then stored again after washing with the chloroform solution. No deactivation was noticed in samples of
product analyzed during the 12-hour run compared to the
samples taken at the start of the test.

PROBLEMS

The most serious problem in operating the pilot plant
was with the inlet strainer on the main feed pump. This
caused a loss of flow to the column pump resulting in the
pumping of air drawn into the recycle line through the column. A check-valve was installed in the line to prevent
its recurrence.

Air introducted into the column in this manner was not
sterile and contaminated the enzyme column. The air was
removed by reversing flow in the column (bottom to top) for
the time necessary to remove bubbles present in the product
stream. No difference was observed in column performance
when flow was reversed. When flow rates were similar, glucose content of the product was the same regardless of flow
direction. The only difference noted when flow was re-
turned to the original downflow was a reduction in the pressure drop across the column to about 50 to 75% of the flow
prior to the upset. The pressure drop increased over a
period of days to the original operating value of 20-30 psi.

During one period of operation several weeks into the
run, it was found that the column temperature could not be
controlled because the cooling water to the process cooler
had eroded the valve. A pressure regulator and a new valve
were installed and remedied the problem.

During the 70 days of operation, the pilot plant was
run on Staley STAR-DRI 24-R most of the time. On two oc-
casions, however, this was not available so Staley STAR-DRI

35-R and Staley STAR-DRI 42-R were substituted to keep the plant operating. The maximum conversion with 35-R was 87.2% glucose, and for 42-R was 86.7% glucose. The reason for the lower conversion is attributable to retrogradation of the hydrolyzed starch and, because these are acid hydrolyzed, the formation products that cannot be broken down by the enzyme. The chromatograms of products of both starches show about 5% material in peaks other than glucose, maltose-isomaltose, maltotriose and starch, which with 5% maltose-isomaltose and 2% starch limits the conversion to glucose.

Tests were run on both of these starches (35-R, 42-R) in a batch immobilized-enzyme reactor and in a batch free-enzyme reactor with the results about the same in both cases, and about equal to the values obtained for the pilot plant starches. The operation of the pilot plant could be monitored by comparing an analysis of the product with the values obtained in the batch runs. Plotting this on an extent of reaction curve shows how the plant is operating.

The analysis of low flow rate product showed a large percentage of maltoses, while the amount of starch decreased slightly and the amount of glucose stayed about the same as the values obtained under high flow rate conditions. Practically, the difference in the product analysis between the low and high flow rates was in D.E., and was primarily affected only by changing maltose concentrations, with the glucose concentration relatively unchanged (Table VI).

Because the conversion in the pilot plant was lower than was desired, a series of tests was conducted on several different hydrolyzed starches to determine if, indeed, the beginning substrate limited the conversion to glucose (Figures 8-10). Batch runs with free enzymes and with immobilized enzymes were conducted using acid thinned Staley 24, 35 and 42 D.E. hydrolyzed starches and G.P.C.'s Maltrins 10,15,20,25 and 42 D.E. syrup - the Maltrins being enzyme thinned, and the syrup acid thinned. In all of these trials the lower D.E. starting material resulted in slightly higher production of glucose in the free enzyme reaction, producing between 1 and 2% higher glucose yields than the immobilized enzyme reaction. The enzyme-thinned substrates (GPC) produced from 2% to 5% higher yields of glucose than the acid-enzyme-thinned starches obtained from Staley (Table VII).

TABLE VI

DATE,FLOWRATE,FEED,TEMP.	G_9	G_8-G_4	G_3	G_2	G	E
*5/6/74, 1020 ml/min 38°C	2.76	3.14	0.91	3.43	89.56	92.4
*5/6/74, 930 ml/min 38°C	2.66	3.29	0.93	3.83	89.29	92.3
*5/6/74, 790 ml/min 38°C	2.32	3.76	0.96	4.40	88.76	92.1
*5/6/74, 700 ml/min 38°C	1.91	3.42	1.01	4.35	89.30	92.5
*5/7/74, 520 ml/min 40°C	1.43	2.55	0.45	4.45	91.11	94.1
*5/8/74, 630 ml/min 40°C	1.73	2.66	0.66	4.46	90.47	93.6
*5/8/74, 735 ml/min 40°C	1.98	2.94	0.78	4.60	89.71	93.0
*5/8/74, 860 ml/min 40°C	2.08	2.91	0.75	4.37	89.89	93.0
*5/8/74, 1020 ml/min 40°C	2.63	2.82	0.78	3.98	89.78	92.8
*5/9/74, 1080 ml/min 38°C	2.97	3.05	0.81	3.51	89.85	92.8
*5/10/74, 480 ml/min 40°C	1.03	1.95	0.28	5.27	91.46	94.6
*5/13/74, 520 ml/min 38°C	1.44	2.92	0.76	4.68	90.19	93.4
*5/15/74, 510 ml/min 38°C	1.44	2.88	0.86	4.71	90.10	93.4
*5/16/74, 275 ml/min 42°C	0.68	2.54	1.04	7.69	88.04	92.7
*5/16/74, 360 ml/min 41°C	0.87	3.04	0.95	6.67	88.46	92.7
*5/22/74, 420 ml/min 38°C	1.18	2.94	0.87	4.97	90.03	93.4
+5/25/74, 420 ml/min 38°C	1.15	4.30	1.69	6.12	86.73	91.3
+5/27/74, 480 ml/min 40°C	1.35	4.56	1.66	5.66	86.76	91.1
+5/27/74, 820 ml/min 40°C	1.93	4.57	1.83	5.24	86.42	90.7
+5/30/74, 520 ml/min 38°C	0.64	5.77	2.31	5.36	86.71	91.1
+6/5/74, 500 ml/min 38°C	1.09	5.42	1.76	5.42	86.30	90.6
#6/10/74, 470 ml/min 39°C	1.24	5.37	1.88	5.70	85.81	90.4
*6/22/74, 470 ml/min 38°C	1.05	2.65	0.54	4.38	91.37	94.3
*6/27/74, 490 ml/min 38°C	1.06	3.02	0.86	4.79	90.27	93.6
*6/27/74, 1030 ml/min 38°C	2.20	3.38	1.10	4.19	89.12	92.4

* S-24R   + S-42R   #S-35R

Tests were conducted with Penick & Ford pearl starch using α-amylase for liquefication and then adding gluco-amylase in a free enzyme batch reaction. This resulted in a 94% yield of glucose. The same procedure for liquefying was then used, but the solution was filtered and reacted in the immobilized enzyme batch reactor resulting in 92.1% glucose. The thinned starch was also run through a contin-uous column (single-pass) resulting, after product filtra-tion, in 93.5% glucose (a glucose content similar to the soluble enzyme). It is believed that by starting with pearl starch, liquefying with α-amylase, then filtering, treating with activated charcoal and running through the immobilized

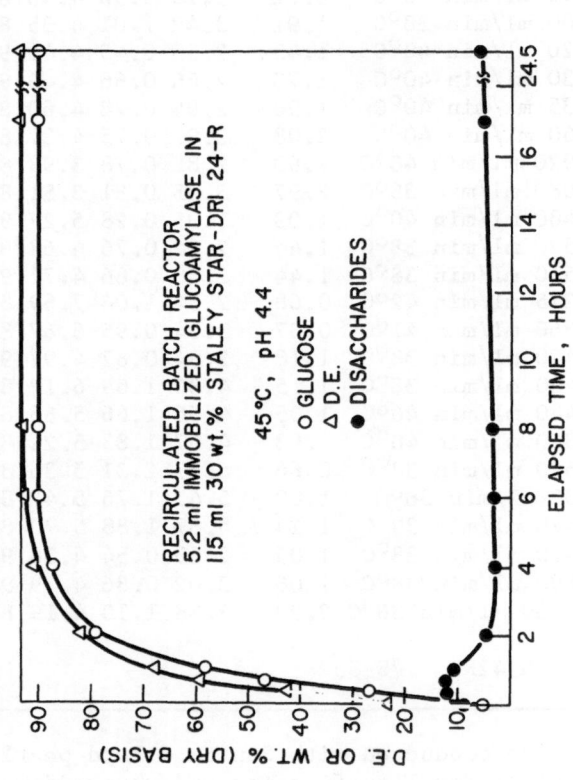

Fig. 8: Formation of Glucose from Staley STAR-DRI 24-R in a Recirculated Batch Reactor with Immobilized Glucoamylase.

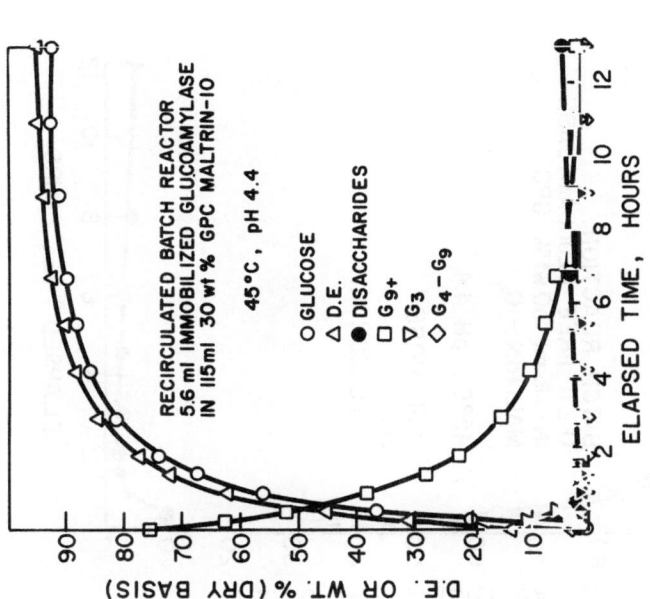

Fig. 9: Formation of Glucose from GPC Maltrin-10 in a Recirculated Batch Reactor with Immobilized Glucoamylase.

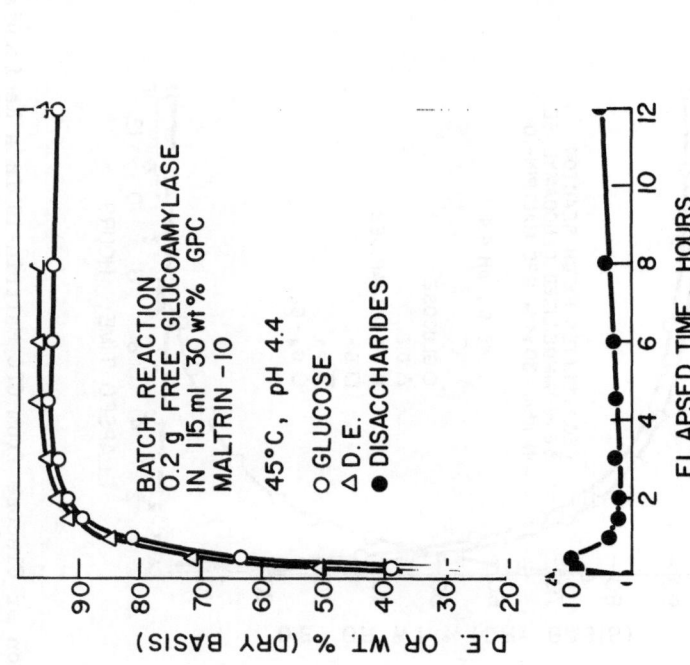

Fig. 10: Formation of Glucose from GPC Maltrin-10 in a Batch Reactor with Free Glucoamylase.

TABLE VII

MAXIMUM GLUCOSE YIELDS

SUBSTRATE DE	MAXIMUM GLUCOSE (WT.%)		MAXIMUM DE (CALCULATED)	
	IMMOB.	SOLUBLE	IMMOB.	SOLUBLE
10	92.8	95.1	95.2	97.1
15	92.0	93.0	94.9	95.8
20	92.2	93.4	94.9	95.8
25	92.8	93.0	95.4	95.4
24	90.5	90.7	93.9	93.8

All starches except the 24 DE material were prepared by enzyme hydrolysis and spray dried.

---

enzyme column in a continuous manner, 93-94% glucose in the product can be obtained with little difficulty.

The general observation that dextrose levels for both immobilized and soluble systems are similar is very encouraging. It is our belief that by using freshly prepared substrate under optimized conditions a 95+ dextrose is possible. In fact, this would be required to make the immobilized GA a commercially viable approach.

Processing Cost Estimate

In order to evaluate the commercial potential of immobilized glucoamylase, an estimate was made of the cost involved in the saccharification of liquefield starch or low D.E. corn syrup. A proposed reactor system for just this hydrolysis step, as shown in Figure 11, was designed to allow maintenance of conversion at a constant level by decreasing flow rate as activity decreases. By utilizing a multiple column system and staggering the column start-up or reloading times, the resulting variation in production rate can be kept within any desired limits. The number of columns necessary to assure a production rate within given limits is a function of these limits and the number of half-lives the IME is used (5). For this system production rate variation limits of $\pm$ 10% were set resulting in at

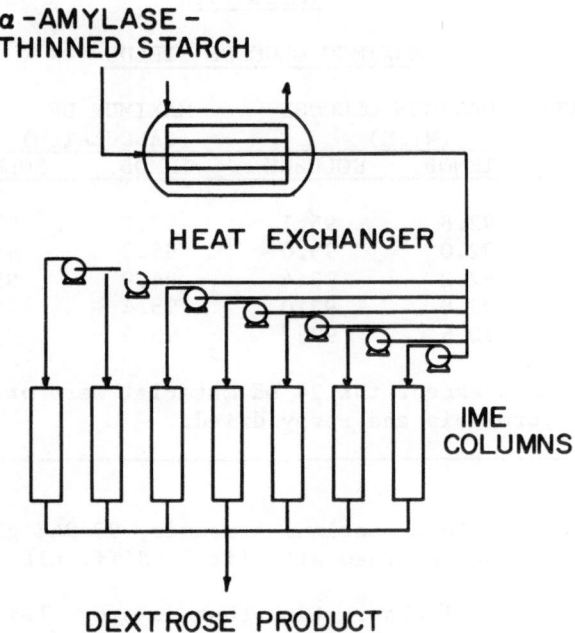

α-AMYLASE-
THINNED STARCH

HEAT EXCHANGER

IME
COLUMNS

DEXTROSE PRODUCT

Fig. 11:   Reactor System.

least seven columns for two half-life IME use. It would
also be desirable to have an additional column available
to provide uninterrupted operation during column loading.

Another possible method of operation would be to maintain conversion at a constant level by raising reactor temperature to increase the reaction rate to compensate for
activity loss.

Cost estimates cover only the saccharification step
(including pH control) and the liquefaction or any subsequent purification steps. Processing costs are given on
the basis of 100-pound (dry weight) quantities of solids
processed.

Column sizes and IME requirements were based on experimentally observed performance. From Table VI the flow rate required in the one cubic foot column for maximum conversion at 40°C can be estimated at 500 ml/min which translates into 16.8 lbs solids per lb IME per day. IME life was estimated from the results for support material C as shown in Table II (13.8 days at 60°C, 51.2 days at 50°C, 402 days projected for 40°C). The relative reaction rates at 40°C, 50°C and 60°C for glucoamylase IME were reported by Havewala and Pitcher (6) to be 1.00, 1.87 and 2.78, respectively.

Equipment costs were estimated from standard literature sources (7,8,9,10) and updated to third-quarter-1974 using the Marshall and Stevens Index. It should be noted that these are preliminary estimates only. However, a contingency of 10%, is included in the final plant cost. Estimated column capacities and plant costs for 100 million lb/yr plants under various operating conditions are shown in Table VIII.

TABLE VIII

PLANT COST ESTIMATES

Basis: 10^8 lb/yr plant

3rd quarter-1974 prices

Operating Temperatures (°C)	Number Half-Lives Enzyme Utilization	Column Capacity (cu.ft.)	Plant Cost ($)
40	3	1247	798,000
40	2	970	634,000
40	1	728	467,000
50	3	668	658,000
50	2	519	512,000
50	1	390	368,000
60	3	448	576,000
60	2	349	442,000
60	1	262	312,000

Labor costs were estimated to be about 11¢/cwt for a plant of this size. Processing costs including labor, IME and capital costs are shown in Table IX. Capital costs are reflected in processing costs by a percentage, 20%, allotted annually for depreciation (10%), maintenance (3 to 5%), insurance (1%) and interest and taxes (4 to 6%).

Since the estimated cost of purified glucoamylase enzyme is less than $2.00 per pound IME, most of the IME cost will be attributable to carrier and immobilization costs. IME costs could reasonably be expected to be in the $5 to $20 per lb range.

The variation of cost with respect to operating temperature, IME cost, number of half-lives enzyme utilization (before columns are re-packed with new IME), and plant capacity was determined. Increased plant size or capacity above 100 million pounds per year would only slightly decrease costs, primarily through labor cost savings. The effect of the other three variables is shown in Table IX. Figure 12 is a contour plot of processing cost as a function of operating temperature and IME cost. From the standpoint of processing cost operation at lower temperatures with 2 or 3 half-life enzyme utilization would be necessary to compete with the 49¢/cwt (12¢/cwt enzyme cost, 11¢/cwt labor cost and 26¢/cwt capital cost) estimated cost of the soluble enzyme process if installed in a new plant.

From Figure 12 it can be seen that at the lower temperatures IME costs as high as $20/lb could be competitive with the soluble batch process, while at 60°C an unrealistically low IME cost of $3/lb would be necessary. The low temperature operation demonstrated in this scale-up is necessary for the economic feasibility of this system.

ACKNOWLEDGEMENTS

We gratefully acknowledge Mr. D.L. Eaton of Corning Glass Works for supplying the porous ceramic carriers and Dr. D.J. Lartigue of Corning Glass Works for his technical assistance in preparing the purified enzyme.

We also acknowledge Novo Enzyme Corp. for the donation of 20 lbs of purified glucoamylase. This enzyme was used for scale-up studies.

TABLE IX

COST OF SACCHARIFICATION BY IMMOBILIZED GLUCOAMYLASE

Basis: 10^8 lb/yr plant (20%/yr plant cost for maintenance, depreciation, interest, taxes, etc.)

Operating Temperature (°C)	Number Half-Lives Enzyme Utilization	IME Cost ($/lb)	Processing Cost (¢/cwt dry solids)			
			5	10	15	20
40	3		32.8	38.7	44.6	50.4
40	2		30.5	37.4	44.2	51.0
40	1		30.6	40.9	51.1	61.4
50	3		37.4	50.6	63.8	77.0
50	2		36.6	52.0	67.5	82.9
50	1		41.5	64.6	87.7	110.8
60	3		63.7	105.0	146.2	187.5
60	2		68.0	116.1	164.3	212.3
60	1		89.5	161.7	233.9	306.9

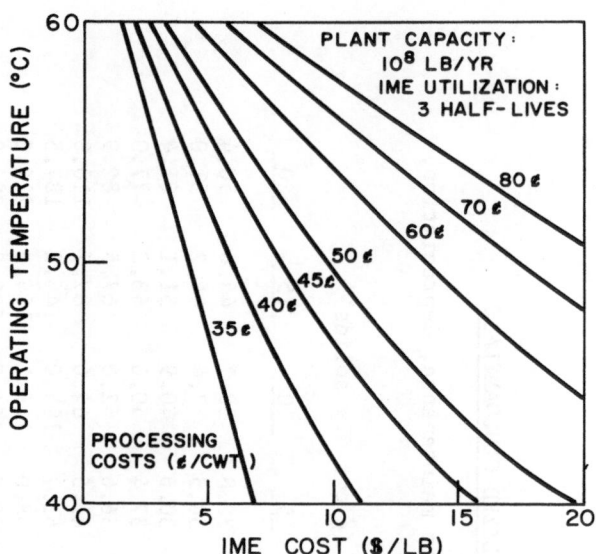

Fig. 12: Processing Cost as a Function of Operating Temperature and IME Cost.

We also acknowledge the excellent technical support
of Mr. R.E. Lindner of Corning Glass Works, who performed
all D.E. determinations and supplied the 60°C column data
for Support B.

The authors gratefully acknowledge support from National Science Foundation grant GI-38101 (ATA73-07783) for the
scale-up studies at Iowa State University.

REFERENCES

1. H.H. Weetall and N.B. Havewala in Enzyme Engineering,
    L.B. Wingard, Jr., Ed., Interscience, New York, 1972,
    pp. 241-266.

2. R.A. Messing, Research/Development, 25, 32 (1974).

3. R.A. Messing in Immobilized Enzymes in Food and Micro-
    bial Processes, A.C. Olson, et al, Ed., pp 129-156(1974).

4. H.H. Weetall, N.B. Havewala, W.H. Pitcher, Jr., C.C. Detar, W.P. Vann and S. Yaverbaum, *Biotechnol. Bioeng.* **16**, 295 (1974).

5. N.B. Havewala and W.H. Pitcher, Jr. in "Enzyme Engineering 2", ed. by E.K. Pye and L.B. Wingard, Jr., 315, Plenum Press, New York (1974).

6. N.B. Havewala and W.H. Pitcher, Jr., "Glucose Production from Cornstarch: Reactor Parameter Study", to be published.

7. H. Popper, "Modern Lost-Engineering Techniques", McGraw-Hill, Inc., New York (1970).

8. J.W. Drew and A.F. Ginder, *Chem. Eng.*, Feb. 9, 100 (1970).

9. B.G. Liptak, *Chem. Eng.*, Sept. 7, 60 (1970).

10. S.P. Marshall and J.L. Brandt, *Chem. Eng.*, August 23, 68 (1971).

FIXED ENZYMES, APPLICATION OF THE STUDY OF PROTEIN STRUCTURE AND FUNCTION [*]

Garfield P. Royer

Department of Biochemistry
The Ohio State University
Columbus, Ohio 43210

Introduction

There are several properties of bound enzymes which suggest their use in the study of protein structure and function. The enhanced stability and convenience of removal permits the use of immobilized proteases in the total hydrolysis and sequencing of polypeptides (1-4). The principal advantage of enzymatic hydrolysis of proteins is that the procedure employed allows the direct determination of labile residues. In conventional acid hydrolysis (110°, 6N HCl, 22 hrs) asparagine, glutamine, and tryptophan are completely destroyed. Glycosides and sulfate, phosphate or carboxylate esters are hydrolyzed. Moreover, as stated by Chin and Wold (4), there could be other labile components of proteins which have never been detected. Soluble enzymes have been used in total hydrolysis (5) but this approach is limited by the problems of autolysis or one enzyme degrading another. Immobilization prevents enzyme-enzyme interaction but allows enzyme-substrate interaction. In sequencing applications the principal advantage is that large amounts of enzyme may be used. Also, the enzyme may be readily separated from the digest. Eventually we hope to develop a sequencing system in which the digest is exposed to a series of bound carboxypeptidases of varying specificity.

When an enzyme is immobilized on a solid support through multiple covalent linkages, it is, in effect, crosslinked. If sufficient bonds are introduced into the enzyme molecule at the proper places, the conformational state of the enzyme should be stabilized. This realization prompted us to devise

299

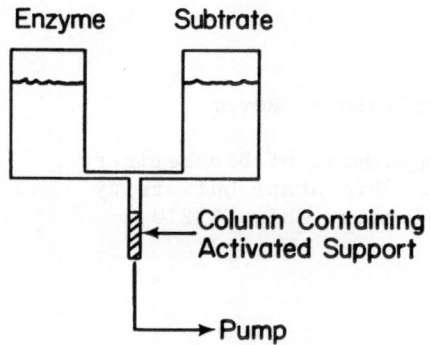

$$dp/dt = k_{cat} E_o = (3.5 \ sec^{-1})(2 \times 10^{-5}) = 7 \times 10^{-5} \ M \ sec^{-1}$$

$$\Delta t_r = 12 \ sec \ \therefore \ \Delta \ p = 0.84 \times 10^{-3} M$$

$$S_o = 5 \times 10^{-3}$$

<u>Fig. 1a</u>:   Schematic representation of the device used for
             covalent bonding of an enzyme to a solid support
             in the presence of saturating levels of a speci-

I. Mixing of Enzyme and Substrate

II Attachment   III Washing

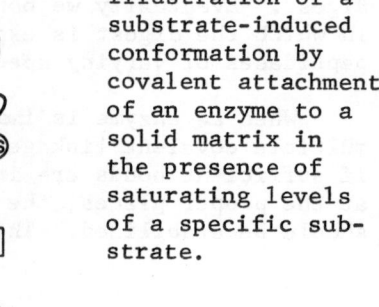

Fig. 1b: Schematic
illustration of a
substrate-induced
conformation by
covalent attachment
of an enzyme to a
solid matrix in
the presence of
saturating levels
of a specific sub-
strate.

the simple apparatus shown in Figures 1a and 1b, which permits immobilization (crosslinking) of an enzyme in the presence of saturating levels of a specific substrate under
such conditions that the enzyme is optimally active. Comparison of the properties of the enzyme forms bound with
and without substrate suggests that a specific substrate
does bring about a change in the conformation of trypsin.

Results and Discussion

Preparation and Characterization of the Immobilized Enzymes.
Pronase and Amino-peptidase M were bound to the arylamine
derivative of porous glass (6,3). Some kinetic properties
of the immobilized derivatives of the respective enzymes
appear in Tables 1 and 2. Pronase is a mixture of proteo-
lytic enzymes isolated from the culture filtrate of S. Gri-
seus (7-9). We showed that the immobilized enzyme retained
considerable activity against two substrates of low molecular weight (Table 1) and two proteins, bovine serum albumin and β-lactoglobulin. The K_m values for the hydrolysis
of the low molecular weight substrates are higher for bound
Pronase which may be attributed to diffusional limitations.
The activation energies and pH optima of bound and soluble
Pronase are similar.

The immobilized aminopeptidase M has the same maximal
activity as the soluble enzyme (Table 2). Also, the activation energies and pH optima of the two enzyme forms are
identical within experimental error. The K_m of the insoluble aminopeptidase M is higher than that of the soluble
enzyme presumably as a result of diffusional limitations.

We have bound three carboxypeptidases to glass (Table
3); carboxypeptidase C was bound to the arylamine derivative
of porous glass by a known procedure (3). This preparation,
although the best of several, has limited activity and stability. Carboxypeptidases A and B were bound to Glycophase
$G^{\#}$ as shown in Figure 2. The coupling reactions were done
at 0-4° in Hepes buffer at pH 7.2, 1 N NaCl for carboxypeptidase A and 0.5 N NaCl for carboxypeptidase B. These
derivatives retained considerable activity against hippuryl-
L-phenylalanine and hippuryl-L-arginine. In both cases 200
mg of insoluble enzyme in a 0.8 cm column would convert
100% of available substrate (1mm) at a flow rate of 2 ml/
min.

	Activity versus				
	BAEE		LPNA		BSA
	$V_m \frac{M}{min}$	$K_m(M)$	$V_m \frac{M}{min}$	$K_m(M)$	$A/\frac{min}{mg}$
Soluble Pronase	1.94×10^{-5}	1.60×10^{-5}	1.00×10^{-4}	0.8×10^{-3}	0.20
Insoluble Pronase	0.52×10^{-5}	5.53×10^{-5}	0.67×10^{-4}	2.0×10^{-3}	0.12
$\dfrac{\text{Parameter for Insoluble Pronase}}{\text{Parameter for Soluble Pronase}}$	0.27	3.46	0.67	2.5	0.57

Table 1: Kinetic parameters for the free and glass-bound forms of Pronase.

	Aminopeptidase M	
	Soluble	Bound
$k_{cat}(S^{-1})$	21 $\pm 0.4^a$	23 $\pm 2^a$
$k_{m(app)}(mM)$	0.28 $\pm 0.01^a$	1.07 $\pm 0.1^a$
$k_{cat}/K_m(S^{-1} \cdot mM^{-1})$	75.0	21.5
$E_a(k_{cal}/mole)$	19.5 $\pm 0.8^b$	17.9 $\pm 1.4^b$
pH optimum	7.5	7.5

^{a} pH 7.5, 25°

^{b} pH 7.5

Table 2. Comparison of the bound and free forms of Aminopeptidase M. The substrate used in these studies
was leucine-p-nitroanilide.

Total Hydrolysis of the aminoethylated A and B chains of
bovine insulin (10) was attempted with immobilized aminopeptidase M (Tables 4 and 5). In the case of the A-chain

Carboxypeptidases bound to glass

Enzyme	Specificity
A	All fast except Asn, Ser, Gly, Glu, Asp, Pro, Arg, Lys
B	Lys and Arg fast all others very slow or not at all.
Y	All released; relative rates on polypeptides unknown.

Table 3. Specificities of the carboxypeptidases which we
have found to porous glass.

Amino acid	Known residues/mole A chain	Recovery
Aminoethylcysteine	4	93
Serine		
Glutamic	2	100
Glycine	1	90
Alanine	1	100
Valine	2	95
Isoleucine	1	90
Leucine	2	95
Tyrosine	2	100
Aspartic Acid	0.4	103
		Mean 96

Table 4:   Recovery of amino acids following the digestion
           of the aminoethylated A-chain of insulin by glass-
           bound aminopeptidase M.

96% digestion occurred. The aspartic acid found is the result of a desamido contaminant found in this preparation of
insulin (Asn-21— Asp 21). The amount of contaminant found
by our enzymatic method agreed well with the value of the
manufactured which was arrived at by electrophoretic studies.
The COOH terminus of the B-chain of insulin has the sequence
of Thr-Pro-Lys-Ala-COOH. Since aminopeptidase M releases
proline slowly or not at all, one might expect low recover-
ies of proline, lysine, and alanine when this enzyme is used
for digestion. This is, in fact, the case as shown in Table
5. Threonine is recovered in low yield indicating that the
X-pro bond is also difficult for this enzyme.

     In the hydrolysis of β-lactoglobulin was used bound
Pronase followed by bound aminopeptidase M. After 2 hrs
approximately 50 spots appear on a peptide map. The number
of spots decreases thereafter as small peptides are reduced
to amino acids. After 6 hrs of digestion with immobilized
Pronase, the extent of hydrolysis is 65% as judged by amino
acid analysis. At this point the digest was transferred to
a column of bound aminopeptidase M. Digestion with this
enzyme produced free amino acids in 93% yield (Table 6).

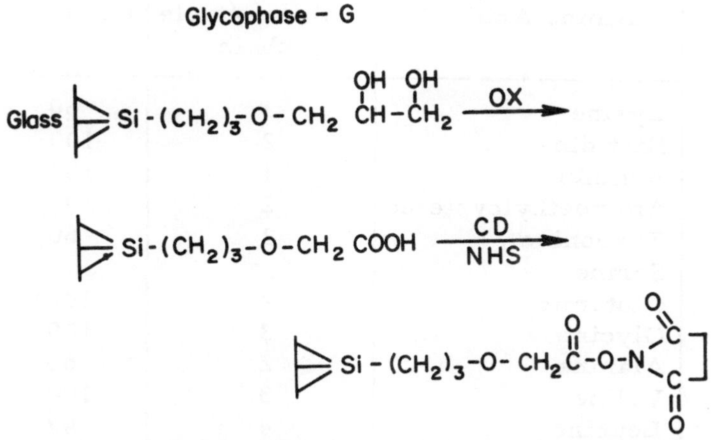

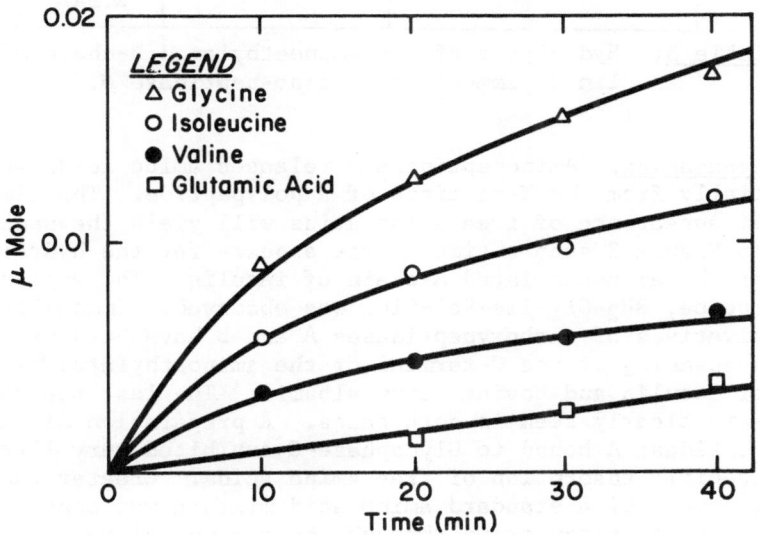

Fig. 2:   A method of attachment of enzymes to Glycophase-G.

Fig. 3:   Time course for the digestion of the N-terminus of
          the aminoethylated B-chain of insulin by aminopep-
          tidase M bound to porous glass.

Amino Acid	Known residues/mole B chain	Recovery
Lysine	1	60
Histidine	2	100
Arginine	1	100
Aminoethylcysteine	2	95
Threonine	1	60
Serine		
Glutamic	2	100
Glycine	3	100
Alanine	2	65
Valine	3	100
Leucine	4	98
Tyrosine	2	105
Phenylalanine	3	103
Proline	1	50
		Mean 87

Table 5: Hydrolysis of the aminoethylated B-chain of insulin by immobilized amino-peptidase M.

Sequencing. Aminopeptidase M releases amino acids sequen-
tially from the N-terminus of a polypeptide. The time course
of appearance of free amino acids will yield the sequence.
In Figure 3 such a time course appears for the hydrolysis
of the aminoethylated A-chain of insulin. The expected sequence, NH_2-Cly-Ile-Val-Glu, was observed. Immobilized derivatives of carboxypeptidases A and B have been used in
sequencing of the C-termini of the aminoethylated β-chain
of insulin and bovine serum albumin. The last two residues
were clearly seen in both cases. A preparation of carboxypeptidase A bound to Glycophase-G exhibited very little non-
specific adsorption of free amino acids. Greater than 90%
recovery of a standard amino acid mixture was observed after
24 hrs of circulation through the enzyme column. The maximum time of digestion with the bound carboxypeptidases was
6 hrs.

Amino acid	Known residues/mole of β-lactoglobulin	Recovery
Tryptophan	2	105
Lysine	15	97
Histidine	2	85
Arginine	3	90
Asparagine		
Threonine	8	95
Serine		
Glutamic		
Glycine	3	113
Alanine	14	105
Valine	10	112
Methionine	4	110
Isoleucine	10	93
Leucine	22	98
Tyrosine	4	98
Phenylalanine	4	98
Proline	8	11
		Mean 93

Table 6: Digestion of β-lactoglobulin with bound Pronase and aminopeptidase M.

Substrate-Induced Conformational Changes

This fundamental question has been approached by a.) study of enzyme conformation in the presence of a specific substrate under conditions where the enzyme is inactive, or b.) study of the enzyme under conditions where the enzyme is active in the presence of non-specific (poor) substrate. Our approach has been to use the apparatus shown in Figure 1 to bind an enzyme to a matrix in the presence of a good substrate, Bz-Arg-OEt$^+$. Trypsin was bonded to CNBr-activated agarose and diazotized arylamino glass in the presence and absence of a specific substrate (ES and E forms respectively). In the latter case the reservoir (Fig. 1) on the right contains only buffer. The bound enzymes were washed thoroughly to remove substrate and adsorbed enzyme. More trypsin was bound to porous glass when the substrate was absent: 5.5% (w/w) for $\underline{E}$ and 3.7% (w/w) for $\underline{ES}$. When agarose was the support the converse occurred: 2.8% for $\underline{E}$ and 4.5% for $\underline{ES}$. These values were arrived at by amino acid analysis. Glass bound trypsin retained 10-15% of the original activity. Agarose-trypsin retained 20-30%.

Trypsin bound to porous glass through the azo linkage showed loss in lysine and tyrosine. The losses are to different extents for the $\underline{E}$ and $\underline{ES}$ forms. The loses of these amino acids most probably occur as shown here:

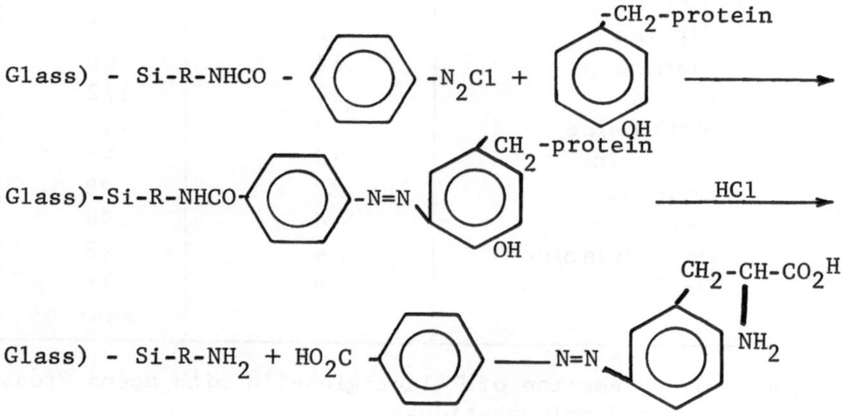

(Reaction I)

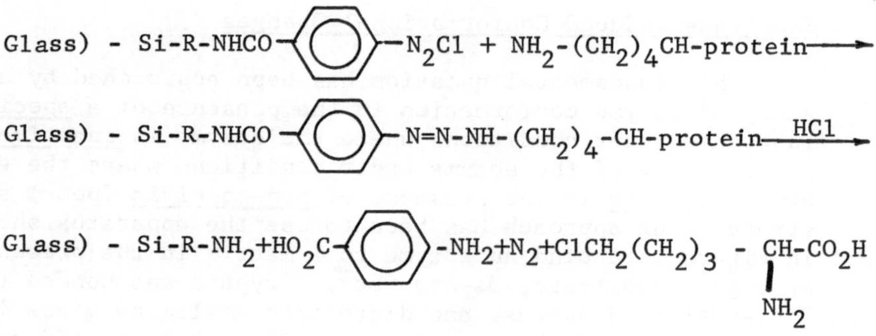

(Reaction II)

With the conventional dual–column mode for amino acid analysis, we observed a peak immediately following norleucine.
This peak is probably ε–chloronorleucine (Reaction II). We
attribute the difference of reactivity between the two enzyme forms to a substrate–induced conformation.   Controls
showed no reaction of Bz⁰-Arg-OEt with the support.
Inagami and coworkers found that the trypsin–catalyzed hydrolysis of acetyl glycine ethylester was accelerated 7 fold
by the presence of methyl guanidine (11). Moreover, they observed an increase in the rate of carboxamido-methylation of
His-46 by iodoacetamide when methyl guanidine was present.
These findings suggested to us that reaction of the ES form
of trypsin with iodoacetamide should be faster than the reaction with the E form.   This is the case (Figure 4).   The
second order rate constant for the modification reaction is
0.26 $M^{-1}hr^{-1}$. Inagami and Hatano (12) found 0.02 $M^{-1}hr^{-1}$ with
soluble trypsin and 0.12 $M^{-1}hr^{-1}$ for soluble trypsin in the
presence of methyl guanidine.

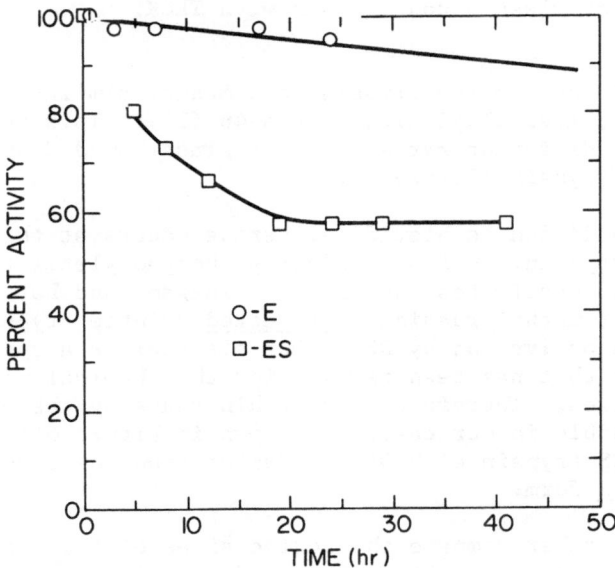

Fig. 4: Reaction of iodoacetamide with glass-bound derivatives
of trypsin. E(o) represents the enzyme which bound in the
absence of substrate and ES(□) represents the enzyme which
was bound in the presence of substrate. Prior to reaction
with iodoacetamide the insoluble enzyme derivatives are
washed thoroughly to remove substrate or loosely bound enzyme.

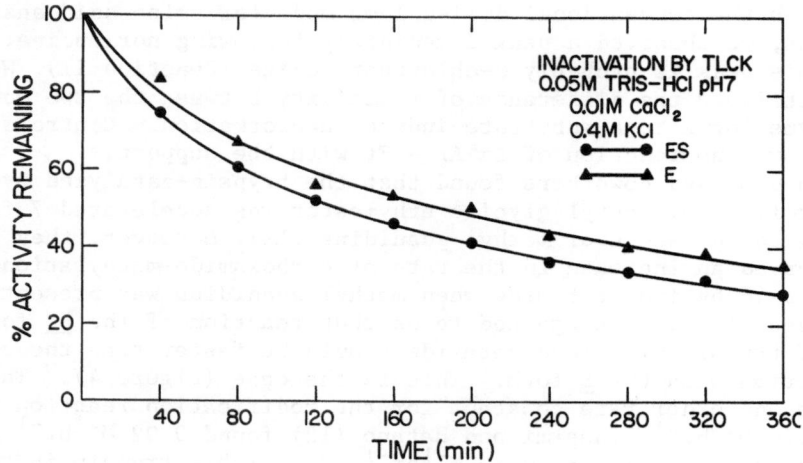

<u>Fig. 5</u>: Time course of the reaction of the <u>E</u> and <u>ES</u> forms
         of glass-bound trypsin with TLCK.

     TLCK inactivates trypsin in a manner similar to iodo-
acetamide, i.e. alkylation of His-46 (13). This reagent
reacts at different rates with the glass-bound <u>E</u> and <u>ES</u>
forms of trypsin (Figure 5).

     In addition to histidine, serine occurs at the active
site of trypsin. DFP specifically phosphorylates this residue and inactivates the enzyme. Inagami and Hatano (12)
found that methyl guanidine <u>protected</u> soluble trypsin
against inactivation by DFP. Our effector is a specific
substrate that has been removed for the chemical modification studies. Therefore, steric hindrance by the effector
is impossible in our case. As shown in Figure 6 the reaction of <u>ES</u> trypsin with DFP is faster than the reaction
with the <u>E</u> form.

     To further compare the active sites of the two enzyme
forms we investigated the pH dependence of the hydrolysis
of Bz-Arg-OEt (Figure 7). The pH optimum of the <u>E</u> form of
Sepharose-trypsin is 8.5 which agrees well with the value
of Knights and Light (14). The pH optimum of the <u>ES</u> form
is 9.0 or 0.5 units higher than the pH optimum of the <u>E</u>
form. These experiments were done in buffer solutions sufficiently concentrated to prevent a large pH gradient between

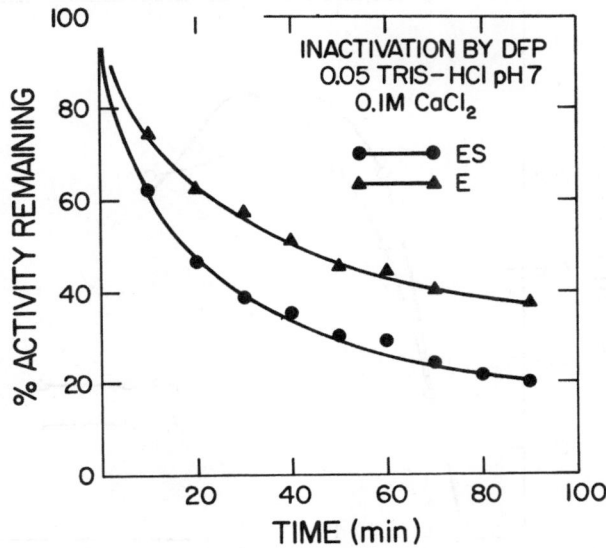

<u>Fig. 6</u>:   Inactivation of the <u>E</u> and <u>ES</u> forms of glass-bound
          trypsin by DFP.

the enzyme matrix and the bulk solution. One explanation
of the shift in pH dependence is that the conformation of
the <u>ES</u> form is such that pK_a's of groups at the active site
are perturbed.

The thermal stability of the <u>E</u> and <u>ES</u> forms of Sepharose-trypsin are also different (Figure 8). Both enzyme
forms were stable up to 60°. A convenient rate of inactivation was produced by incubation at 70°. At timed intervals the samples were cooled to 25° and assayed with Bz-Arg-
OEt as substrate. It is clear (Figure 8) that the <u>ES</u> form
is more stable than the <u>E</u> form. The inactivations occur in
a biphasic manner in which both phases may be treated as
first-order reactions. The rate constants for the fast re-
action are 40×10^{-3} min^{-1} for <u>E</u> and 25×10^{-3} min^{-1} for <u>ES</u>.
For the slow reaction the rate constants are 12×10^{-3} min^{-1}
for <u>E</u> and 4×10^{-3} min^{-1} for <u>ES</u>. It is quite reasonable to
suggest that the <u>ES</u> is more stable as a result of a greater

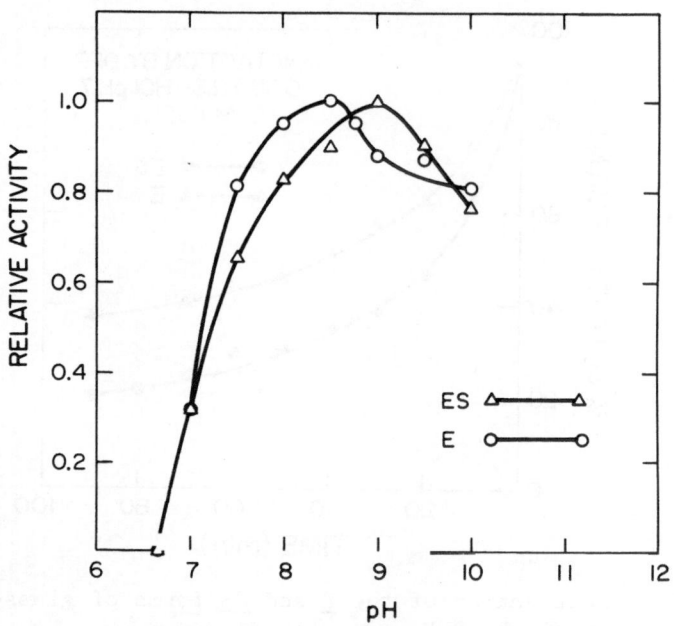

<u>Fig. 7</u>:   The pH dependence of the hydrolysis of Bz-Arg-OEt
          catalyzed by the <u>E</u> and <u>ES</u> forms of agarose-bound
          trypsin.

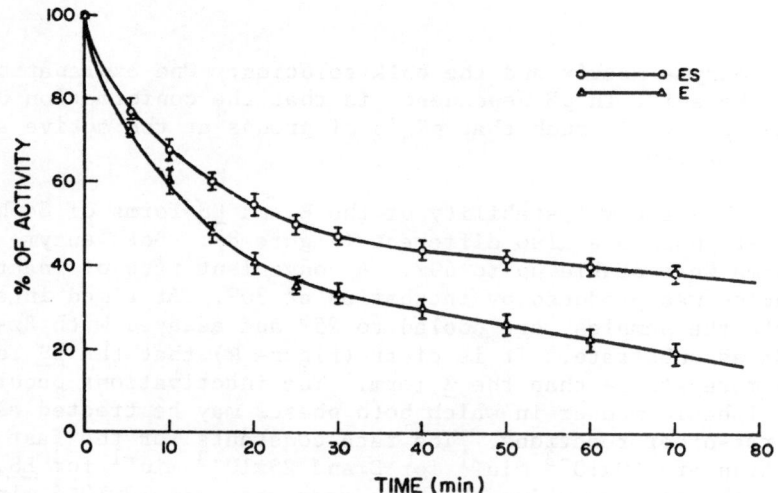

<u>Fig. 8</u>:   Denaturation of the two forms of agarose-bound tryp-
          sin by incubation at 70°.

number of points of attachment between enzyme and matrix.
This cannot be observed directly since no change in amino
acid analysis is observed for the bound enzyme as compared
to the soluble enzyme. For the case of ES on glass a
greater number of points of attachment for the ES as compared with E was observed directly (Table 7).

In the presence of soluble trypsin the hydrolysis of
Bz-Arg-OEt and Bz-Arg-NH$_2$ follows Michaelis-Menten kinetics.
Hydrolysis of these substrates by Sepharose-bound trypsin
also follows Michaelis-Menten kinetics; apparent K_m values
appear in Table 8. The K_m for the hydrolysis of Bz-Arg-OEt
is considerable higher for the matrix-bound enzyme. For
hydrolysis of Bz-Arg-NH$_2$. The K_m values are similar for
the bound and free forms.

Our results on the attachment of trypsin to a solid
support in the presence of a specific substrate are consistent with the induced fit theory of Koskland (15-18). This
theory states that a specific substrate binds to the enzyme
and brings about a conformational change. The result of
the change is the repositioning of catalytic groups to a
favorable arrangement. A poor substrate may bind well, but
not in such a way as to induce the rearrangement of groups
at the active site. Our observations of changes in the reactivity of groups on the enzyme, changes in pH dependence,
changes in thermal stability and differences in the Michaelis constants of the two enzyme forms support the theory.

Residue	Residues Found		
	Enzyme Form		
	Native	E	ES
Lys	14.0	7.7	6.3
Tyr	10.0	5.7	3.6
Bonds/ Enzyme Molecule		10.6	14.1

Table 7. Amino acid analysis of the E and ES forms of
          glass-bound trypsin.

	E	ES	Native
Bz-Arg-OEt	0.41 ± 0.02	0.24 ± 0.01	0.004^a
Bz-Arg-NH$_2$	2.4 ± 0.2	1.6 ± 0.1	3.3^b 6.8^c

[a]Value of Baines, N.J., Baird, J. B., and Elmore, D. T. (1964) Biochem. J. 90, 470-476.

[b]Chevallier, J. and Yon, J. (1966) Biochem. Biophys. Acta 122, 116-121 (pH 7.9, 35⁰).

[c]Goldstein, L., Levin, Y., and Katchalski, E. (1964) Biochemistry 3, 1913-1919 (pH 7.6, 25⁰).

Table 8. Michaelis constants for free trypsin and agarose-
          bound E and ES forms.

There are practical aspects of enzyme immobilization in the presence of a specific substrate. First, the substrate appears to protect the enzyme during the binding process. Secondly, if the "induced" form of the enzyme is trapped by immobilization the substrate specificity should be expanded. In other words an enzyme bound in the presence of a good substrate should then be a more effective catalyst in reactions which involve poor substrates. We have observed more extensive digestion of a polypeptide substrate by the ES form of trypsin than by the E form. The study was done by peptide mapping and suggests less specific cleavages take place in the presence of the ES form. Finally, our coupling technique in which the support is held stationary is appropriate for support materials which tend to fracture with stirring.

Footnotes

* This work was supported by grants from the National Science Foundation (GI-32141), the National Institutes of Health (GM-19507) and the Pierce Chemical Co. The author is also indebted to the National Science Foundation and the Japan

Society for the Promotion of Science.

Glycophase-G is a trademark of Corning Glass Works. The N-hydroxy-succinimide derivative was kindly provided by Corning. Glycophase-G is now marketed by the Pierce Chemical Co., Rockford, Illinois.

‡ Abbreviations used: Bz-Arg-OEt, N-α-benzoyl-L-arginine ethylester; TLCK, N-p-toluenesulfonyl-L-lysine chloromethyl ketone; DFP, diisopropylphosphorofluoridate.

References

1. Royer, G.P. and Andrews, J.P., *Polymer Preprints*, 13, 848 (1972).

2. Royer, G.P. and Andrews, J.P., *J. Macromol. Sci.*, A7 (5), 1167 (1973).

3. Royer, G.P. and Andrews, J.P., *J. Biol. Chem.*, 248, 1807 (1973).

4. Chin, C.C.Q. and Wold, F., *Anal. Biochem.* 61, 379 (1974).

5. Hill, R.L., *J. Biol. Chem.* 237, 63 (1962).

6. Royer, G.P. and Green, G.M., *Biochem. Biophys. Res. Commun.* 44, 426 (1971).

7. Namoto, M., Narahashai, Y. and Murakami, M., *J. Biochem.* 48, 906 (1960).

8. Morihara, K., Tsusuki, H. and Oka, T., *Arch. Biochem. Biophys.* 123, 572 (1968).

9. Trap, M. and Birk, Y., *Biochem. J.* 116, 19 (1970).

10. Humbel, R.E., Deron, R. and Neumann, P., *Biochemistry* 7, 621 (1968).

11. Inagami, T. and Murachi, T., *J. Biol. Chem.* 239, 1395 (1964).

12. Inagami, T. and Hatano, H., *J. Biol. Chem.* 244, 1176 (1969).

13. Shaw, E. and Springhorn, S., Biochem. Biophys. Res. Commun. 27, 391 (1967).

14. Knights, R.J. and Light, A., Arch. Biochem. Biophys. 160, 377 (1974).

15. Koshland, D.E. Jr., Proc. Nat. Acad. Sci. USA 44, 98 (1958).

16. Thoma, J.A. and Koshland, D.E. Jr., J. Am. Chem. Soc. 82, 3329 (1960).

17. Koshland, D.E. Jr., Advan. Enzymol. 22, 45 (1970).

18. Koshland, D.E. Jr., Ann. Rev. Biochem. 37, 359 (1968).